Economics of New Food Product Development

other AVI books

Amerine, Berg and Cruess	Technology of Wine Making, 2nd Edition
Brink and Kritchevsky	Dairy Lipids and Lipid Metabolism
Charm	Fundamentals of Food Engineering, 2nd Edition
Copson	Microwave Heating
Desrosier	Attack on Starvation
Desrosier	The Technology of Food Preservation, 3rd Edition
Desrosier and Rosenstock	Radiation Technology in Food, Agriculture and Biology
Esmay	Principles of Animal Environment
Goldblith and Joslyn	Milestones in Nutrition
Graham	Safety of Foods
Greig	Economics of Food Processing
Hall *et al.*	Encyclopedia of Food Engineering
Hall and Hedrick	Drying of Milk and Milk Products, 2nd Edition
Heid and Joslyn	Fundamentals of Food Processing Operations
Henderson	The Fluid-milk Industry, 3rd Edition
Inglett	Corn: Culture, Processing, Products
Joslyn and Heid	Food Processing Operations, Vols. 1, 2, and 3
Kazarian	Work Analysis and Design
Kramer and Twigg	Fundamentals of Quality Control, 3rd Edition, Vol. 1
Lawrie	Proteins as Human Food
Levie	Meat Handbook, 3rd Edition
Matz	Cereal Science
Matz	Cereal Technology
Peterson and Tressler	Food Technology the World Over, Vols. 1 and 2
Potter	Food Science
Sacharow and Griffin	Food Packaging
Schultz	Food Enzymes
Schultz and Anglemier	Proteins and Their Reactions
Schultz, Cain and Wrolstad	Carbohydrates and Their Roles
Schultz, Day and Libbey	Chemistry and Physiology of Flavors
Schultz, Day and Sinnhuber	Lipids and Their Oxidation
Smith	Pototoes: Production, Storing, Processing
Sperti	Probiotics
Thorner and Herzberg	Food Beverage Service Handbook
Tressler, Van Arsdel and Copley	Freezing Preservation of Foods, 4th Edition, Vols. 1, 2, 3, and 4
Tressler and Joslyn	Fruit and Vegetable Juice Processing Technology, 2nd Edition
Van Arsdel	Food Dehydration, Vol. 1
Van Arsdel and Copley	Food Dehydration, Vol. 2
Webb and Johnson	Fundamentals of Dairy Chemistry
Webb and Whittier	Byproducts from Milk, 2nd Edition
Weiser	Practical Food Microbiology and Technology, 2nd Edition
Weiss	Food Oils and Their Uses
Woolrich	Handbook of Refrigerating Engineering, 4th Edition, Vols. 1 and 2
Woolrich and Hallowell	Cold and Freezer Storage Manual

Economics of New Food Product Development

by NORMAN W. DESROSIER, Ph.D.

Director of Research,
NABISCO Research Center,
Fair Lawn, New Jersey

and JOHN N. DESROSIER

Pelham Manor, New York

WESTPORT, CONNECTICUT

THE AVI PUBLISHING COMPANY, INC.

1971

THE AVI PUBLISHING COMPANY, INC.
Westport, Connecticut

REGISTERED AT STATIONER'S HALL
LONDON, ENGLAND, 1971

Library of Congress Catalog Card Number: 76-156797
ISBN-0-87055-102-7

Printed in the United States of America
BY MACK PRINTING COMPANY, EASTON, PENNSYLVANIA

Preface

The idea for this book came while organizing and holding a "Meeting in Miniature" on new product development for the New York Section of the Institute of Food Technologists earlier this year. It was a conference dealing solely with new product development, with the specific focus on trying to build better understanding between industrial research laboratory people and manufacturing, marketing, financial, and advertising groups, with an underlying hope of uncovering ways of improving all of our performances in the subject area.

The conference brought together representatives of small and large R&D systems, production organizations, marketing organizations, advertising agencies, new product companies, diversified and single product companies, and selected governmental laboratories and institutions. There were food scientists and technologists, manufacturing experts, market research specialists, financial and economic experts, legal representatives, media specialists, communications experts, representatives of the behavioral sciences, and business management specialists from various levels of responsibilities in many organizations.

It was clear at the outset and confirmed as a result of the conference that there was much which needed to be done if a consensus were to be drawn describing the path to success in new product development. A consensus existed; however, it had to be extracted from what had been written and said at the meeting and combined with supporting materials of an auxiliary nature outside the scope of the meeting to bring balance and understanding to the subject.

We have tried to do this within this volume. What we believe emerges is the system, which, in minor variation here and there, is being used and proven successful in many companies and marketplaces in various countries where it is being tried. The system is designed to yield successful new mass-marketed consumer products in an expeditious manner.

The system evolved from a rapidly unfolding body of knowledge in the fields involved. In fact, sufficient experience has been accumulated with this system that it is now possible to write a course of instruction for students of the subject who would like to gain the insights needed to participate effectively in this exciting activity. This is our basic purpose here.

The method used in this book is to explore new product development and describe it as a recurrent phenomenon that can be understood, organized, oriented, and managed.

There have been many contributions in one form or another to this book and we have been most encouraged by the cooperation of each. Also, where possible, we have included selected abstracts of contributors and participants in the "Meeting in Miniature," as well as abstracts of contributions drawn from other sources as needed. We have acknowledged specific contributions, for there are many, as they appear in the text.

However, we have made specific and substantial extractions from certain published works deemed essential to the presentation. In many instances, a sequence of Figures or Tables or examples were needed to develop a point effectively. While we make no claim of authorship of these, and we have identified them clearly in the text, we wish to make more specific acknowledgement here of our gratitude and to offer our most sincere appreciation to each for allowing us to employ their published works in this manner.

Their publications contained large units of the vital components required for a meaningful, expert overall treatment of the subject. These contributors, in alphabetical order, and by subject, are:

Anonymous contributors to government publications which represent enormous effort, and which we have drawn upon with no means of acknowledging each personally; Lee Adler, for information on systems approaches to marketing and the splendid Tables and Figures which capture so clearly the essence of the matter; T. S. Angelus, for the background data on new product failures and suggestions for improving the situation; Jack Baen, for the detailed description of the elements of market research; L. W. Bass, for his analysis and examples of innovations in industry; Bart Bates, for his description of venture organizations; W. M. Bright, for his case history and description of specific mergers and acquisitions; R. J. Bouthilet, for his examples and description of market forecasting techniques; Robert Buckles, for legal guidance in protecting ideas and inventions; G. D. Carson, for his examples of how to use the Bayesian Theory in the food field; P. A. Cavalier, for describing in elemental terms and figures the financial functions in a company; R. H. Cotton, for the case history of how a large diversified corporation manages new product development; J. G. Crockett, for his views and examples of proving out new product ideas prior to laboratory activity; F. M. Depew, for examples of legal issues in new product work; R. K. Diehl, for indicating how to segment a market and identify products of potential interests, with diagrams; S. G. DeNisco, for insights and operating procedures used by advertising agencies; P. F. Drucker, for his descriptions of organizations, functions, and their management; S. R. Goodman, for making available his financial decision guide as applied to new products; T. J. Gordon, for insights into the business climate of 1985; A. E. Haase, for the Figures and Tables concerning the case history of Space Sticks; J. C. Kable for the examples of changes in the food industry; R. J. Lavidge, for his examples on equating market opportunities with new product potentials; R. E. LeBleu, for the data projections of business in 1985; T. Levitt, for the background and examples on innovative imitation; F. A. Jenkins, for the material on communications; B. J. F. Lonergan, for his examples of human understanding; M. L. Liston, for material on creative people and their organizational needs; G. E. MacDougal, for the studies and tables on mergers; J. J. Moder, for Tables and Figures on CPM and PERT; J. F. Magee, for the information on the logistics of distribution; L. D. McGlauchlin, for the material on long-range planning; F. V. Malek, for the insights into preparing for merger negotiations; R. E. Morse, for the information on product life cycles; W. H. O'Meara, for the Tables and material on selecting profitable new products; A. D. Orman, for the information on using systematic approaches in selecting new product ideas; the late L. B. Parsons, for his views and insights into the total R&D area; C. R. Phillips, for information and figures on CPM and PERT; E. E. Rice, for the analysis showing changes in the American dietary habits; E. R. Seawell, for the examples and material on using the computer in merchandizing; M. A. Straus, for the Tables and Figures and insights into research people and organizations; C. B. Riter, for the case history on building a business from technology; W. H. Sly, for the Tables and data on R&D management and computer applications; and, H. K. Weiss, for the information and Tables on market forecasting.

We also wish to express our appreciation to Mr. Edward C. Borsk, Editor, Harvard Business Review; Mr. Richard D. McCormick, Editor, Food Product Development; and Mr. Karl O. Herz, formerly Editor, Food Technology, for permission to incorporate copyrighted materials from their publications in this manuscript. The specific areas are noted in the text.

A special note of appreciation is due to Mr. Lee S. Bickmore and NABISCO, Inc., for encouragement and the opportunity to pursue the subject.

To all who have generously given of their time and thought in this undertaking, we are most genuinely obliged. However, errors of commission and omission in thought and/or presentation are ours alone to bear.

A special note of thanks is due to Dr. Donald K. Tressler and Mr. John J. O'Neil of Avi Publishing Company for their patience and encouragement in bringing out this book.

It is also a distinct pleasure to acknowledge the editorial assistance given to us by Ann, Nancy, James and Brian Desrosier, and the suggestions and review of the financial presentations by Prof. James R. McCrisaken, Indiana State University, Terre Haute, Ind.

Norman W. Desrosier
John N. Desrosier

November 1970 *Pelham Manor, New York*

Contents

SECTION I

Nature of the Changing Business Scene

CHAPTER 1

Introduction

Basic to our understanding of new product development is our belief that it requires human cooperation. Due to the nature of the task, it calls forth teamwork and a division of labor. It indicates the areas within which teamwork is divisible. At the same time, it invites men to specialize in these particular areas of knowledge and encourages their timely contributions to the particular task, adding momentum to the flow of accomplishments. Concurrently, the task calls forth some order, some procedure, for setting guidelines for the production of consumer goods and the generation of capital. The task calls forth some method, some system, for establishing how much of a particular kind of product is to be produced. It requires some mechanisms for assigning work functions and calls for order in distributing production. And, in the end, the task calls forth common sense—an understanding of things in their relationship to us and others. Such is the realm of new product development.

Common sense is workable and people working effectively together share it. They seek knowledge, not for the sake of pleasure of contemplation, but to use knowledge in making and doing. By making and doing together, men breed and maintain enormous structures of technology, economics, and politics. Furthermore, common sense also adds a series of new levels or dimensions in the network of human relationships.

It is clear that the development of common sense involves a change, not only in men to whom things are related, but also to things which are related to them.

The ability to focus experience and common sense to the task at hand is intelligence at work. Intelligence also manifests itself in works of art. Thus, human intelligence is not only practical but artistic.

An example of this was shown by Lonergan (1957). He suggested that in primitive cultures, hunger was linked to eating by rather simple bodily movements and sequences. But, primitive hunters took time out from hunting to make spears, primitive fishermen took time out from fishing to make nets, and primitive farmers took time to fashion plows. Neither the spears, nets or plows were in themselves necessarily objects of the desire to eat at the outset. Still, with notable ingenuity and effort these implements were created and often were adorned, where possible, with artistic designs.

The intervention of intelligence revealed to them that desires to eat were recurrent, work to provide food was recurrent, and the comparatively brief time spent making spears or nets or plows was amply compensated for by the greater ease with which more game, fish, or crops could be harvested, gathered, or generated on an infinite series of occasions. Artistic designs, such as on spears, were probably individual expressions, useful in distinguishing, for example, one spear from another, revealing further the nature of creativity.

Such an intervention of intelligence is itself recurrent. As products of human ingenuity, spears, nets, and plows illustrate not only the idea of the old mechanical arts but the notion of the foundation of modern technology as well. As pieces of material equipment, the same objects are initial examples of the concept of capital formation. Lonergan believes that the history of man's material progress lies essentially in the expansion of these ideas.

To carry the concept further, he noted that as inventions accumulate, they set opportunities calling for more inventions. The new inventions complement the old to suggest further improvements, and to reveal fresh possibilities and eventually call forth, in turn, a succession of technologically higher viewpoints, each built on the preceding, that mark the epochs in man's material progress viewed in history. Moreover, he noted, this advance of practical intelligence is registered not merely in men's minds and later in books, but more obviously in concrete products.

In correspondence with each advance, there is generally a technological obsolescence of capital equipment. New models produce better goods more efficiently, and vigorous economic growth calls forth efficiency.

As technology stimulates an economy, so the economy stimulates human institutions and organizations which leads to the next notion: most men get ideas and ideas are what start the process. But, these ideas reside in different minds and the different minds do not quite agree at the outset. Of itself, communication only reveals the differences. What is wanted is persuasion: and the most effective persuader becomes the leader or chief. One can see the problem of effective agreement as recurrent. Each step in the process of development of human institutions is an occasion on which minds differ. New insights must be communicated. Enthusiasm has to be initiated. A common decision must be reached (Lonergan 1957).

Beyond the common sense of the laborer, the technician, and the entrepreneur, there is the managerial specialization of common sense. Its task is to provide the catalyst that brings men of common sense together. Management often has to deal with incomplete accumulations of data to be complemented and modified by the further understandings that arise from the situation at hand. While it involves understanding of industry and commerce, its special field is dealing with men. It has to discern when to press for full performance or when to compromise, when delay is wisdom or when it spells disaster, when widespread consent must be gained or when action must be taken in spite of opposition. It has to command attention, win confidence, set forth concretely the essentials of a case, make its own decisions, and secure the agreement of others.

It is indeed fascinating to consider that, as in the field of physics so in the field of human events and relationships, there also appear to be classical laws that combine concretely. Human actions are recurrent; their recurrence is regular; and the regularity is the functioning of patterned sets of relations that yield conclusions.

Inventions outlive their inventors and the memory of their origins. Capital has utility, not in itself, but in the acceleration it imparts to the stream of useful things. The management machinery of agreement and decision is permanent, or practically so, yet it is also a self-adapting locus for an endless series of agreements and decisions.

ORGANIZATIONS AND PEOPLE

All human institutions are organizations. They are complex and multidimensional and have functional, moral, and political dimensions. They do not exist for their own sake; they are means to ends; and the test of performance always resides in results achieved outside themselves (Drucker 1968).

Nevertheless, an organization must decide what results it wants to achieve. It is obviously not possible to manage an organization unless this is established and agreed upon. The specific contribution it expects to be paid for must be clearly identified, he found.

An organizational structure cannot be established without first deciding what is to be done, and how it is to measure whether it is doing it or not. There is usually agreement on what should be done, and disagreement on what should be done first.

There is always a conflict in organizations between those wanting to do better what is already being done and those who want to do different things. It is never possible to give final answers to these matters; situations are constantly changing, according to Drucker.

A large organization cannot be versatile; it can do only a small number of tasks at one time. Large organizations are effective through mass action rather than through agility. Yet, mass action enables a great many kinds of knowledge to be combined for specific tasks through many people.

Organizations separate those people who decide from those who do not. Organization is thus a maximizing device. It can maximize the human resources of knowledge, Drucker believes.

Further, an organization requires the ability to stop doing unproductive work, and to slough off the obsolete. In the process it can take on new efforts. What one should not do is the first order decision that needs to be made. Then, priority is given the tasks at hand.

Organizational inertia always pushes for continuing what it is doing. The organization is always in danger of being overwhelmed by yesterday's tasks and problems and being rendered sterile by them (Servan-Schreiber 1969).

Voluminous literature suggests that organizational management of all kinds has certain features in common: (A) They all need balance between the needs of the organization and the needs and desires of individuals in them. (B) Each requires a structure to determine tasks and demands. (C) Each needs to balance the need for order versus the need for flexibility and individual activity. (D) Each needs rules of organization. Unless each person has decision-making power at each level of organization, there will be little or no performance. Balance is needed between authority in situations and knowledge of individuals. A dynamic coexistence is required. (E) Each organization must measure results (Drucker 1968).

There is conflict in all large organizations between efficiency, or emphasis on order, and effectiveness, or emphasis on results. For convenience, call efficiency "first" and effectiveness "second." The following are generally found: (A) In the first, effort is seen as the central task. In the second, results are the issue. (B) In the first, order in the organization is paramount. In the second, creativity and vitality are sought. (C) In the first, administration is the strength of organization. In the second, administration is a support, a necessary evil, which prevents collapse of the organization, hence, must be kept to the minimum level possible. (D) In the first, attempts are made to make mediocrity capable of producing predictable results and is based on what proponents believe is a realistic view of man and organizations. In the second, it tries to liberate creativity and believes in its presence, which is also a realistic view.

Actually, a balance of both points of view is needed; organizations must tend to be efficient and effective (Galbraith 1967).

In operational matters relating to the business of new product development, the issue that needs attention is personal effectiveness of people within an organization. Drucker (1968) suggests an organization is a legal fiction. By itself, an organization does nothing. Individuals decide and plan and do.

Yet, organization actually is something people need. Organization is useful to people; to some, it becomes a "tool" and, at the same time, produces results needed by the society and the community (Lodge 1970).

Organization needs the right decisions to get results. Since present day organization executives are generally not informed directly by the environment (i.e., they are not in direct contact with people who own or are being served by the organization), the executive must ultimately decide certain matters by himself or he will not achieve. This places new demands on people. While common sense principles work in an organization, initially they reveal mainly what structure the people involved require, what kind of person is needed to operate an organization, and what he must do (Drucker 1968).

While policy objectives (i.e., the means of measuring results, the means of measuring personnel effectiveness, and the specific management tasks involved) are all different in different organizations, they nevertheless all operate in the same field of human endeavors; all have the same dimensions of organization to deal with; and, all deal with making organizations functional. Yet, does it not all seem to be just common sense?

Clearly, schemes of recurrence exist and function. They can thus be understood. They are to be explored below in specific reference to new food product development in these terms and in its various facets.

GUIDE TO NEW PRODUCT DEVELOPMENT

Thus, it seems reasonable to address ourselves to the task of uncovering the recurrent new product phenomenon, and to set guidelines for it. We plan to do this by first focusing on pertinent background information and then on the nature of the structural order of a system which obviously yields successful new products in the sophisticated consumer food product markets, and in an apparently endless stream. We believe that a guide to new product development is contained in the chapters to follow.

Since the food industry exists in a field of changing markets, insights into these changes eventually lead to understanding the business of new products. The changing scene is discussed in Chap. 2. In Chap. 3, we examine how companies analyze the changing scene and forecast basic markets of specific interest. In Chap. 4, the subject of generating new product ideas to meet these market opportunities is explored in depth. In Chap. 5, we investigate the market research required to be brought to bear by the system as it relates to consumers.

In Chap. 6, we explore the communications needed in mass marketing companies, and the means of organizing them for this purpose. In Chap. 7, we present the mechanisms with which large industrial organizations, starting with a raw idea and/or tested concepts, create prototype products, bring the new product physically into being and then into mass production. In Chap. 8, we examine the organization and function of industrial research laboratories in new product development and their management to maximize results. In Chap. 9, the nature of the creative environment needed is examined with particular reference to the innovation of new products in industry and the research climate demanded by it.

In Chap. 10, we explore the use of financial and business decision guides in new product development; and in Chap. 11, some critical consumer and legal aspects of the undertaking. In Chap. 12, modern merchandising is examined, with particular reference to the situation generated, given a fully operating merchandising program and organization, by the introduction of new products to the product mix.

In Chap. 13, we address ourselves to trying to improve the success potential in new product development; and in Chap. 14, this is extended to product and market exploitations, and how these are combined and brought to bear on acquisitions and mergers of companies.

In Chap. 15, the new venture concept is examined as it is being applied in building new businesses. This leads into another kind of consideration, dealing with what to do with a new product which a company develops and then decides that it cannot pursue.

It was not many years ago that the food industry was composed of a large number of small independent companies, each producing its own special product line, for specific geographic markets. The revolution in convenient-to-use food products and rising consumerism were relatively slow to become apparent. In Chap. 16, these events are viewed, including a forecast of what we believe lies ahead for new product development in the food industry.

BIBLIOGRAPHY

de Chardin, P. T. 1955. The Phenomenon of Man. Harper and Row, New York.

Drucker, P. F. 1968. The Age of Discontinuity. Harper and Row, New York.

Eckler, A. R. 1970. Profit from 1970 census data. Harvard Business. Rev. *48,* No. 4, 4-7.

Galbraith, J. K. 1967. The New Industrial State. Houghton Mifflin Co., Boston.

Lodge, G. C. 1970. Engines of Change. Alfred A. Knopf, New York.

Lodge, G. C. 1970. Top priority–renovating our ideology. Harvard Business Rev. *48,* No. 5, 43-55.

Lonegran, B. F. 1957. Insight. Philosophical Library, New York.

McGarrah, R. E. 1970. Swords into plowshares. Harvard Business Rev. *48,* No. 4, 36-40.

McLuhan, M. 1964. Understanding Media. McGraw Hill Book Co., New York.

Servan-Schreiber, J. J. 1969. The American Challenge. Atheneum Publishers, New York.

CHAPTER 2

Changing Business Scene

Insight into new product development is hinged on insights into human needs, be they personal, group, community, country, or civilization. In new product development, it is well known that that which possesses a high probability of success in one period, community, country, or civilization may possess little or no probability in another. The differences may lie only in a minor way in outward and material factors and almost entirely in the set of understandings that are operating in each situation. These sets of understandings are accessible, persuasive, and potentially usable. In this chapter, we plan to explore the changing business scene, with specific reference to the food industry. The natural extension of the material presented to other industries is suggested to be highly probable.

PHENOMENON OF CHANGE

Just as the individual selects his own course of action from the range of possibilities confronting him, so too does the group, the community, the organization, the institution. Commonly available understandings are communicated to one and another and, by persuasion, these modify the mentalities of the people involved. With these, they determine the course of events to follow out of the alternatives presented by the changing scene.

As was shown earlier for individual persons, so now with organizations. They, too, have common sense and it, too, is an incomplete set of understandings ever to be completed, according to Lonergan (1957), and differently in each set of situations. Their adaptation is continuous and too rapid for them ever to be able to stand fixed in some rigid set of definitions and propositions.

The common sense of an organization or group is continuously being adapted. As understanding grows, common sense of the group grows. A group is not outfitted with it like armor. Rather, the growth of understanding is more like the germination, division, and differentiation of cells in seeds, shoots, and plants.

So, too, the communities of man. Those that exist and function know only a story of their beginnings. Men have a bare outline of their development and of their complexity.

Faced with such complexity, to understand the workings of a dynamic social system of men, we must inquire from many men in their many walks of life. If we were successful at this, we might discover the functional unity that organically binds men together in the endlessly varied units of an enormous jigsaw puzzle.

Lonergan suggests a duality to be grasped in this unity. First, as intelligent beings, man is the sponsor of the order imposed by common sense, even though each person is but a partial holder of the truth. Initially and almost spontaneously, he identifies himself with the good associated with satisfying desires. These desires are not solely animal impulse or ego satisfying. Man is an artist at work, be it ever so humble. He is also practical. His practicality is part of his pursuit of dignified living. His aim is not just raw and isolated satisfaction. While he may never even dream of giving up food and water, still he does look for a succession of varied acquisitions and attainments. If he never forgets his personal interests, common sense tells him he is a member of one or more groups. Man is a social being and he soon learns that the elemental basis of community is not found in the raw pursuit of self interests but in the spontaneous occurrence of the benefit of living in harmony with others. A sense of belonging then provides the dynamic notion for common enterprise of many forms.

The discoveries of intelligence of past centuries may appear to some to have been only incidental additions to the fabric of human living, compared to man's present capability to occupy even the moon. But, discover they did—seed and plant, bread and wine, wheel and carriage, paper and book, pump and power, and the process continues. Discoveries continue to penetrate and overwhelm nearly every aspect of human living.

Technology and capital formation also continue to grow. In successful business enterprises, there is a succession of new technology and capital formation, and it is planned. It is also recurrent and, hence, can be subjected to intelligent inquiry. The voyage to gain the insights is believed worthy of the effort required.

TABLE 2.1

EXPECTED TRENDS IN U.S. ECONOMY

	1950	1960	1970	1980	1985
1. GNP (in billions of 1958 dollars)	380	415	790	1100	1300
2. Output per manhour (1957-1959 = 100)	69	112	143	157	168
3. Median annual earnings of salaried engineers and technical men (in thousand $)	6	7.5	12	16.5	20
4. Gross weekly hours of production and nonsupervisory workers	40	37.5	37.5	35	34
5. Purchasing power of the consumer dollar (1957-1959 = $1.00)	1.2	1.0	0.8	0.65	0.55
6. Gross average hourly earnings of production and nonsupervisory workers	1.3	2.2	3.0	4.0	4.5
7. Unemployment rates					
Nonwhite (percentage of increase)	10.9	11.1	7.7	6.0	5.7
White (percentage of increase)	6.8	5.0	3.9	4.0	4.1

Source: Gordon and LeBleu (1970).

SCENARIO FOR 1985

If the projections shown in Tables 2.1 and 2.2 are anywhere near correct, industry can expect real changes in markets in the next two decades. The gross national product (GNP), output per man hour, salaries and wages, service industries, state and local governments, education investments, women

TABLE 2.2

EXPECTED TRENDS IN U.S. LABOR FORCE

	1950	1960	1970	1980	1985
Employees in selected industries	(% of all employed persons)				
Manufacturing	14.5	24.5	24.0	23.0	22.0
Wholesale and retail	15.0	18.0	18.5	19.0	19.3
Services	10.0	13.0	17.0	19.0	21.8
State and local government	8.5	10.5	13.0	17.0	17.7
Federal government	3.0	3.0	3.8	5.0	5.8
Employed persons 16 yr and older, by occupation					
White-collar	40.0	43.0	48.0	52.0	54.0
Blue-collar	40.0	37.0	37.5	33.0	30.5
Services	12.0	12.0	17.5	20.5	26.0
Farm	12.0	8.0	3.5	3.0	2.3
Median years of school completed, by labor force	(Civilians over 18 yr)				
Professional and managerial workers	12.0	13.0	14.4	16.6	17.2
Clerical and sales workers	12.0	11.9	12.0	12.8	13.0
All workers	8.0	10.3	12.3	12.6	12.8
Craftsman, operatives, and laborers	8.0	8.8	10.8	11.8	12.3
Employed nonwhite persons 16 yr and older by occupation	(% of all employed persons)				
Total	8.9	10.2	10.7	11.4	12.0
Blue-collar	3.5	4.3	4.8	4.9	4.9
Services	3.4	3.3	3.4	3.4	3.6
White-collar	0.5	1.7	2.1	3.0	3.3
Farm	1.5	0.8	0.4	0.2	0.2
Reserves, private pension, and deferred profit-sharing plans	(In billion $)				
	20	53	140	230	275
Education expenditures under public programs	(In billion $)				
Total	8	17	42	76	96
State and local	5	14	29	40	36
Federal	3	3	13	36	60
Employed females 14 yr and older by occupation	(% of all employed persons)				
Total	30.0	34.5	37.0	39.0	42.5
White-collar	13.0	17.5	21.5	22.5	23.0
Services	7.5	8.0	8.0	8.7	11.0
Blue-collar	6.5	6.5	5.5	6.0	7.0
Farm	3.0	2.5	2.0	1.8	1.5
Average life expectancy at birth	(Years)				
	66.5	67.5	67.8	71.2	73.8
U.S. private, long-term, direct investment abroad	(In billion $)				
	15	32	80	125	175
Union members as a proportion of labor force	(Percentage)				
(1930–6%, 1950–4.5%)	21.5	24.0	21.7	22.0	23.5
Social insurance expenditures under public programs	(In billion $)				
Total	9	22	45	76	95
State and local	6	6	7	6	6
Federal	3	16	38	70	89

working, life expectancy, and social insurance will continue their steady climb. Work hours per week, purchasing power of the consumer dollar, unemployment rates, blue collar and farm work forces will continue to decline. Together, these mean real changes in the food industry are ahead. The trends are already available for review.

NATURE OF CHANGE IN THE FOOD INDUSTRY

A casual analysis of the national averages of total food consumption in the United States reveals that there has not been much change per person since 1900 (Fig. 2.1), and this can be extrapolated to the year 1985 without great risk.

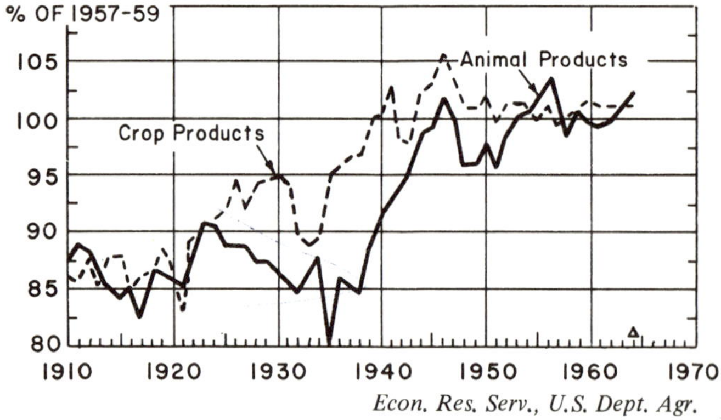

FIG. 2.1. FOOD CONSUMPTION PER CAPITA FOR CROP AND ANIMAL PRODUCTS

Similarly, if we examine specific food consumption (Table 2.3) we see a familiar picture. The most consistent pattern is that of decline in the use of potatoes and cereals. Dairy products, fruits, and vegetables increased until the 1940-1950 decade, but then declined slightly although remaining above the 1900 levels. Meat, poultry, and fish products declined slowly until about 1930, remained steady for a few years, and then increased to the present level, substantially above that of 1900. Eggs show a varying pattern.

Such generalizations, while giving a broad picture, conceal many significant trends. For example, they do not demonstrate the use of processed versus unprocessed foodstuffs, nor do they distinguish one product from another within a general category. All studies of specific changes in the American dietary, however, have found many implications for the future. Some of these patterns of change also demonstrate the nature of the new types of foods continuing to be developed to meet changing consumer needs.

COMMODITY CHANGES

Analysis of available statistics for food consumption in the United States reveals significant trends, however. The findings of Rice (1967), Diehl (1969), and Hase and Riter (1970) contain the following (see Fig. 2.2 and 2.3):

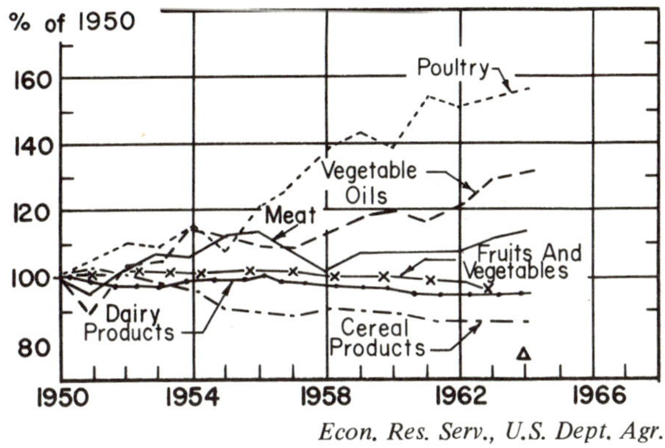

FIG. 2.2. FOOD CONSUMPTION PER CAPITA

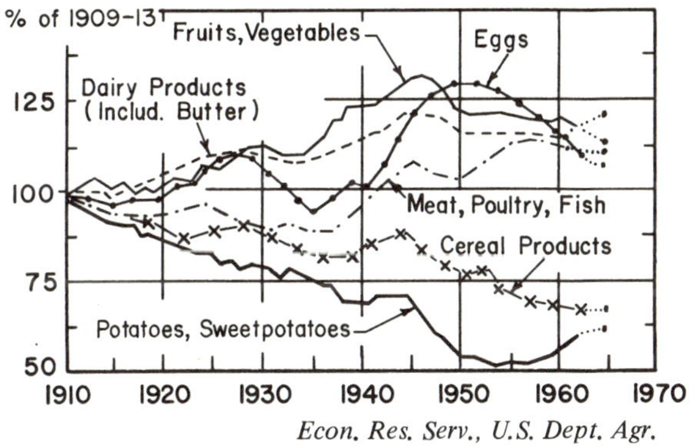

FIG. 2.3. TRENDS IN OUR EATING HABITS: PER CAPITA CIVILIAN CONSUMPTION USING CONSTANT RETAIL PRICES AS INDEX WEIGHTS

Dairy Products

The total consumption of milk has increased from 94 billion pounds in 1920 to about 125 billion pounds in 1970. Total per capita use of dairy

TABLE 2.3

PER CAPITA POUNDS OF FOOD CONSUMPTION RETAIL WEIGHT EQUIVALENT, 1909-1963[1]

	Meat, Poultry, Fish					Dairy Products		Fats and Oils				Fruits			
									Including Butter[4]						
Year	Meat[2]	Poultry	Fish[3]	Total	Eggs	Including Butter	Excluding Butter	Excluding Butter	Animal	Vegetable	Total	Fresh	Processed	Total	Melons[5]
1909	147.7	17.2	(13.0)	177.9	35.5	378	360	(23.2)	(31.3)	(9.7)	(41.0)	123.3	7.2	130.5	40.0
1910	139.0	18.0	(13.2)	170.2	37.1	355	337	(23.6)	(31.8)	(10.1)	(41.9)	123.0	7.6	130.6	39.0
1911	144.6	18.1	(13.3)	176.0	39.9	344	325	(22.7)	(31.7)	(9.6)	(41.3)	139.1	8.4	147.5	38.0
1912	139.1	17.4	(13.3)	169.8	37.7	386	369	(22.2)	29.2	(9.6)	(38.8)	142.4	9.2	151.6	39.0
1913	136.8	16.9	(13.5)	167.2	36.8	376	360	(23.7)	27.4	(12.8)	(40.2)	118.9	8.2	127.1	38.0
1914	133.0	16.9	(13.7)	163.6	35.8	359	342	(26.5)	29.4	(14.1)	(43.5)	146.3	9.9	156.2	35.0
1915	129.5	16.8	(13.2)	159.5	37.9	357	340	(26.0)	29.1	(14.1)	(43.2)	140.7	11.2	151.9	36.0
1916	134.5	16.2	(13.0)	163.7	36.3	354	337	(24.9)	30.7	(11.5)	(42.2)	122.1	12.6	134.7	36.0
1917	129.3	15.6	(12.9)	157.8	34.1	363	347	(24.4)	28.1	(12.0)	(40.1)	118.5	14.3	132.8	36.0
1918	134.9	15.6	(12.9)	163.4	34.4	390	376	(29.3)	28.3	(15.1)	(43.4)	109.1	12.4	121.5	37.0
1919	133.7	16.6	(13.6)	163.9	36.8	369	354	(28.1)	28.1	(15.2)	(43.3)	111.9	16.9	128.8	36.4
1920	130.5	16.0	(13.8)	160.3	36.3	378	363	(24.8)	28.4	(11.3)	(39.7)	130.3	16.7	147.0	42.5
1921	128.3	15.7	(12.5)	156.5	36.4	378	362	23.2	28.5	11.0	39.5	103.7	14.0	117.7	45.2
1922	131.8	16.5	(13.3)	161.6	38.3	377	360	26.0	32.2	10.9	43.1	132.2	14.3	146.5	46.2
1923	141.4	16.8	(12.7)	170.9	39.6	366	348	26.9	34.0	10.7	44.7	132.3	14.8	147.1	38.1
1924	141.0	15.9	(13.0)	169.9	39.3	376	358	27.7	33.8	11.7	45.5	135.6	15.4	151.0	42.9
1925	133.8	16.4	(13.1)	163.3	38.6	377	359	28.7	32.2	14.6	46.8	121.2	17.8	139.0	41.8
1926	131.1	16.3	(13.4)	160.8	41.0	377	359	29.4	32.4	15.3	47.7	147.1	18.4	165.5	43.5
1927	128.7	17.4	(14.2)	160.3	41.5	376	358	29.6	32.7	15.2	47.9	116.0	19.5	135.5	38.6
1928	126.1	16.7	(14.1)	156.9	41.0	376	358	30.1	32.4	15.3	47.7	133.7	19.4	153.1	38.4
1929	125.3	16.4	13.9	155.6	40.5	381	363	31.1	32.0	16.7	48.7	128.1	18.5	146.6	40.7
1930	123.3	17.9	12.2	153.4	40.2	378	360	31.2	31.7	17.1	48.8	119.3	19.1	138.4	41.5
1931	125.2	16.2	10.8	152.2	40.4	376	358	30.0	33.5	14.8	48.3	147.0	16.4	163.4	41.3
1932	125.8	16.6	10.4	152.8	38.0	379	360	28.3	33.9	12.9	46.8	115.9	16.6	132.5	36.4
1933	130.0	17.3	10.7	158.0	36.0	377	359	28.7	33.3	13.6	46.9	114.8	18.0	132.8	35.8
1934	135.7	15.9	11.2	162.8	35.0	368	349	29.9	33.1	15.4	48.5	115.0	18.6	133.6	36.1

1935	110.9	15.4	12.5	138.8	34.0	372	354	30.5	29.1	19.0	48.1	134.2	21.8	156.0	37.5
1936	122.6	16.4	13.7	152.7	35.0	374	357	32.7	30.2	19.3	49.5	124.1	25.2	149.3	36.8
1937	119.7	16.4	13.8	149.9	37.3	377	360	32.5	28.9	20.4	49.3	142.7	24.3	167.0	38.9
1938	120.3	15.5	12.8	148.6	37.6	376	359	32.5	29.4	19.7	49.1	131.0	26.5	157.5	37.5
1939	126.8	17.1	12.7	156.6	37.9	381	364	32.9	31.7	18.6	50.3	148.6	29.5	178.1	35.9
1940	135.9	17.5	13.0	166.4	38.7	382	365	33.2	33.0	17.2	50.2	138.4	33.6	172.0	36.8
1941	137.0	18.8	13.2	169.0	37.7	385	369	35.1	32.0	19.2	51.2	146.8	31.9	178.7	34.1
1942	135.1	21.1	10.7	166.9	38.1	409	393	32.6	31.3	17.2	48.5	129.3	31.5	160.8	31.4
1943	143.3	26.1	9.9	179.3	41.6	427	415	33.2	26.7	18.3	45.0	116.0	27.0	143.0	30.8
1944	150.9	23.5	10.7	185.1	42.5	436	424	32.0	25.8	18.1	43.9	138.2	27.8	166.0	36.2
1945	140.9	25.5	11.9	178.3	48.4	451	440	31.1	24.0	18.0	42.0	133.4	33.7	167.1	37.7
1946	147.8	23.5	12.8	184.1	45.5	447	436	32.3	23.3	19.5	42.8	131.6	47.8	179.4	38.4
1947	147.4	22.1	12.3	181.3	46.8	424	413	33.9	25.7	19.4	45.1	139.6	40.8	180.4	35.1
1948	138.4	21.8	13.1	173.3	47.8	407	397	35.7	24.4	21.3	45.7	127.9	42.8	170.7	34.5
1949	137.6	23.3	12.9	173.8	47.4	406	396	35.2	24.1	21.6	45.7	121.1	42.4	163.5	33.2
1950	137.8	25.1	13.8	176.7	48.5	407	396	38.4	25.1	24.0	49.1	106.7	43.8	150.5	30.4
1951	132.4	26.5	13.2	172.1	49.3	410	400	35.6	24.1	21.1	45.2	114.9	42.9	157.8	31.6
1952	139.4	27.2	13.3	179.9	49.2	412	403	38.7	23.0	24.3	47.3	109.4	45.2	154.6	30.2
1953	146.4	27.1	13.6	187.1	47.8	407	398	38.7	22.6	24.6	47.2	104.2	45.8	150.0	32.5
1954	144.9	28.5	13.5	186.9	47.4	404	395	39.9	21.5	27.3	48.8	100.4	45.6	146.0	32.3
1955	152.2	26.7	12.9	191.8	46.9	407	398	40.2	22.7	26.5	49.2	94.6	48.8	143.4	33.0
1956	154.9	30.0	12.9	197.8	46.6	409	400	39.8	23.1	25.4	48.5	94.8	49.5	144.3	30.1
1957	146.7	31.8	12.8	191.3	45.9	402	394	39.3	21.8	25.8	47.6	92.4	51.2	143.6	27.6
1958	139.6	34.4	13.3	187.3	44.9	396	388	40.4	21.7	27.0	48.7	89.9	50.0	139.9	29.4
1959	146.6	35.6	13.7	195.9	44.7	391	383	41.7	21.5	28.1	49.6	93.5	49.4	142.9	26.9
1960	146.7	34.7	13.2	194.6	42.4	385	378	41.0	20.1	28.4	48.5	90.6	51.3	141.9	28.0
1961	145.5	38.0	13.7	197.2	41.3	376	369	41.0	20.8	27.6	48.4	86.4	50.5	136.9	27.3
1962	147.1	37.4	13.6	198.1	41.0	375	368	41.5	20.1	28.7	48.8	82.1	51.3	133.4	25.0
1963[11]	151.7	38.0	13.6	203.3	40.0	375	368	43.0	19.4	30.5	49.9	73.7	50.0	123.7	26.4

Source: U.S. Dept. of Agr.

[1] Retail weight data comprise the quantity data used in computing the per capita food consumption index. Final consumer products that are derived from a combination of primary food groups, such as bakery products, are measured and reported in the form of their primary ingredients, such as flour, shortening, and eggs. Civilian consumption only, beginning 1941.

[2] Includes game and edible offal.

[3] Includes 2.0 lb per capita of game fish 1909-1951, increasing 0.1 lb annually to 3.0 lb in 1961.

[4] Includes product weight of butter and margarine; other items in terms of fat content.

(Continued)

Table 2.3 (*Continued*)

Fruits Melons, Baby Food	Vegetables					Vegetables, Soup, Baby Food[6]	Potatoes and Sweet Potatoes[5]	Beans, Peas, Nuts, Soya Products[5,7]	Flour and Cereal Products[8]	Sugars and Other Sweeteners[9]	Coffee Tea, Cocoa	All Foods[10]			Year
		Processed													
	Fresh[5]	Canned	Frozen	Total	Total[5]							Animal Products	Crop Products	Total	
170.5	(189)	15.3	–	15.3	(204.3)	204.5	211.9	16.6	300	85.6	9.8	605	1012	1617	09
169.6	(188)	14.5	–	14.5	(202.5)	202.9	221.1	16.3	295	87.9	9.6	576	1016	1592	10
185.5	(184)	15.6	–	15.6	(199.6)	200.1	181.0	16.0	291	90.2	9.1	573	986	1559	11
190.6	(187)	18.7	–	18.7	(205.7)	206.4	201.5	16.4	286	88.6	11.3	606	1014	1620	12
165.1	(180)	19.8	–	19.8	(199.8)	200.7	209.4	16.0	283	93.0	9.7	591	993	1584	13
191.2	(184)	19.2	–	19.2	(203.2)	204.3	178.3	15.2	280	92.3	9.9	571	989	1560	14
187.9	(181)	18.0	–	18.0	(199.0)	200.1	206.0	15.0	278	88.9	11.3	567	1004	1571	15
170.7	(182)	16.1	–	16.1	(198.1)	199.2	166.3	14.3	277	88.6	12.4	568	943	1511	16
168.8	(182)	18.9	–	18.9	(200.9)	202.2	171.3	18.2	268	90.0	14.1	566	947	1513	17
158.5	(185)	22.3	–	22.3	(207.3)	208.6	196.3	16.6	260	90.9	12.0	602	961	1563	18
165.2	180.5	21.3	–	21.3	201.8	203.1	177.9	16.3	259	105.9	13.1	583	958	1541	19
189.5	195.9	18.2	–	18.2	214.1	215.5	165.8	14.4	242	102.3	12.7	588	956	1544	20
162.9	185.3	16.6	–	16.6	201.9	203.4	178.9	13.8	229	101.2	12.8	583	916	1499	21
192.7	190.2	16.8	–	16.8	207.0	209.2	167.3	13.8	242	119.8	13.0	592	972	1564	22
185.2	183.9	21.2	–	21.2	205.1	208.0	192.4	15.4	241	106.1	14.1	594	976	1570	23
193.9	189.9	22.7	–	22.7	212.6	215.0	166.6	16.5	239	116.1	13.4	601	975	1576	24
180.8	190.2	25.4	–	25.4	215.6	217.4	169.5	16.1	235	118.8	12.3	593	968	1561	25
209.0	190.0	25.7	–	25.7	215.7	217.7	145.3	16.3	238	120.0	14.1	593	979	1572	26
174.1	189.4	22.1	–	22.1	211.5	213.7	159.7	17.7	239	119.2	13.5	593	956	1549	27
191.5	188.0	22.8	–	22.8	210.8	213.3	161.6	17.6	239	121.1	13.0	588	976	1564	28
187.3	194.0	25.7	–	25.7	219.7	222.6	174.2	16.8	236	113.9	14.1	591	985	1576	29
179.9	196.7	28.3	–	28.3	225.0	227.2	145.4	17.6	228	125.7	13.6	585	958	1543	30
204.7	192.6	25.2	–	25.2	217.8	219.5	151.1	18.3	226	115.3	14.3	584	967	1551	31
168.9	198.0	22.0	–	22.0	220.0	221.6	156.3	16.4	223	110.0	13.8	586	926	1512	32
168.6	194.5	21.9	–	21.9	216.4	217.9	151.3	15.9	213	110.9	14.4	586	908	1494	33
169.7	200.2	23.2	–	23.2	223.4	225.8	154.6	17.7	204	110.2	13.6	580	914	1494	34

193.5	198.2	26.1	–	26.1	224.3	226.9	161.9	18.1	204	110.6	15.7	556	953	1509	35
186.1	198.1	27.6	–	27.6	225.7	228.9	145.6	19.1	208	111.5	16.1	576	938	1514	36
206.0	196.8	29.3	0.4	29.7	226.5	230.4	143.4	18.0	203	109.1	15.0	576	951	1527	37
195.1	198.9	31.0	0.4	31.4	230.3	234.5	146.0	19.5	204	107.9	16.0	575	946	1521	38
214.1	198.3	31.7	0.5	32.2	230.5	235.4	138.1	19.7	201	113.3	16.9	590	960	1550	39
208.9	197.9	34.2	0.6	34.8	232.7	238.0	136.1	19.5	199	108.1	17.4	603	947	1550	40
212.9	192.0	36.6	0.7	37.3	229.3	234.6	142.7	19.2	199	117.9	18.0	608	967	1575	41
192.6	199.5	39.5	1.1	40.6	240.1	245.7	142.3	22.5	201	102.8	15.1	629	941	1570	42
174.2	204.4	36.7	0.7	37.4	241.8	246.0	142.4	20.5	208	100.5	13.8	663	925	1588	43
202.8	211,8	34.1	1.6	35.7	247.5	252.7	152.4	20.1	190	108.8	16.3	677	962	1639	44
205.7	217.5	42.9	1.9	44.8	262.3	267.9	136.3	20.4	201	93.2	16.9	691	961	1652	45
218.8	201.3	46.0	2.0	48.0	249.3	255.4	136.9	19.2	192	93.6	20.8	691	958	1649	46
216.5	190.0	39.8	2.5	42.3	232.3	238.4	135.5	16.4	173	113.7	18.6	667	930	1597	47
206.4	186.8	37.4	2.9	40.3	227.1	233.8	113.8	17.4	170	106.4	19.1	643	888	1531	48
197.9	174.9	38.3	2.9	41.2	216.1	222.7	118.8	16.7	169	108.5	19.5	641	874	1515	49
182.2	170.3	41.2	3.2	44.4	214.7	221.7	114.2	19.4	167	113.1	17.6	647	858	1505	50
190.7	165.1	41.5	4.0	45.5	210.6	217.7	117.2	17.3	165	106.3	17.3	644	850	1494	51
186.2	161.8	40.8	4.9	45.7	207.5	215.2	105.9	17.2	162	108.4	17.3	655	835	1490	52
184.0	157.7	42.2	5.1	47.3	205.0	213.0	112.1	16.2	158	108.1	17.3	655	836	1491	53
179.8	155.4	41.1	5.4	46.5	201.9	210.1	111.6	16.3	155	106.0	14.7	652	822	1474	54
177.8	154.6	42.2	5.9	48.1	202.7	211.1	113.7	16.3	152	106.3	15.0	660	820	1480	55
175.8	154.4	42.7	6.0	48.7	203.1	211.7	107.2	17.1	150	107.7	15.5	668	813	1481	56
172.9	152.8	42.7	6.2	48.9	201.7	210.6	112.6	16.4	148	104.3	15.3	654	808	1462	57
171.1	149.2	43.2	6.6	49.8	199.0	208.5	107.3	16.3	149	107.4	14.7	642	805	1447	58
171.8	147.3	43.2	6.8	50.0	197.3	207.1	109.7	16.9	148	107.8	15.2	645	807	1452	59
172.0	149.9	43.0	7.0	50.0	199.9	210.0	109.5	16.2	147	108.8	15.1	635	810	1445	60
166.4	147.8	43.0	7.1	50.1	197.9	208.3	110.0	16.9	146	109.1	15.4	628	802	1430	61
160.6	144.8	44.4	7.7	52.1	196.9	207.4	108.8	16.6	144	109.8	15.5	627	795	1422	62
152.4	145.1	45.1	7.4	52.5	197.6	208.2	111.4	16.7	144	110.0	15.6	631	789	1420	63

5 Data for melons, fresh vegetables, potatoes, sweet potatoes, and dry beans and peas include consumption of home garden produce.
6 Excludes estimated quantities of meat, poultry, and fish base soups duplicated elsewhere.
7 Since 1958 approximations for soya products are continued at 1.3 lb.
8 Corn sugar and syrups are with sugars and other sweeteners.
9 Excludes sugar used in production of canned and frozen fruit, canned fruit juices, canned vegetables, and unskimmed sweetened condensed milk.
10 Includes spices and herbs with approximations prior to 1918.
11 Preliminary.

products expressed as whole milk has varied from a low of 628 lb (1960) to a high of 838 lb (1930), with a general downward trend since 1930.

Processed products within the dairy classification have been used in increasing amounts. During this period the consumption of low fat milks of all types, skim milk, buttermilk, 2% milks, etc., doubled, going from 15.6 to 34.2 lb per person. Fluid milk usage remained relatively constant. The consumption of butter declined steadily, yet the consumption of spreads has not decreased. The decline in butter has been offset by an increase in margarine consumption. The intake of cheese has increased slightly. Ice milks have had a large increase in consumption, while ice cream has shown little gain. The ice milks represent about $^1/_4$ of the frozen desserts sold in 1970.

Grains and Flour

From 1900 to 1960, overall consumption of flour declined from 214 lb to 116 lb per capita; during this period the use of wheat flour in processed foods increased from 55 to 85 lb, representing a change from 26 to 72% of the total flour consumption. In 1900, wheat accounted for 72% of all cereal foods sold; this changed gradually to 81% in 1960, largely at the expense of corn. The use of rye and buckwheat flours also declined from 5.8 to 1.2 lb, contrary to what one might assume. Per capita consumption of wheat-based ready-to-eat breakfast cereals has declined from 3.2 lb in 1930 to a present value of about 2.6 lb per capita. All ready-to-eat cereals total about 4.7 lb per capita or about 3% of the total food usage of cereals (Rice 1967).

The decline in the use of bread accounts for most of the decreased use of cereals. The decline in the use of corn meal is a second factor. Flour mixes, developed during the last 3 decades, now account for 8-9% of the total cereal volume.

Potatoes

Since 1900, potato usage has declined from 187 lb to less than a 100 lb per person. Canned, frozen, and dehydrated potato products simultaneously became important commodities. In 1950, processed potato products amounted to about 1 lb per person; by 1960, they had increased to over 10 lb per person and amounted to 10% of the potato usage.

Animal and Fish Products

The long-term increase in the consumption of meat, poultry, and fish is concealed in trends for individual products. Since 1900, veal, lamb, turkey, and fish have appeared as minor contributors to the total animal protein product group, despite the fact that each is an important commodity. Their total volume now is less than 15% of the meat sold, about the amount that chicken has captured. There has been an increasing use of turkey, about double over the last 20 yr. Since 1950, use of beef has increased by more than 50%. Retail cuts of beef now supply about 1/3 the total meat, fish, and poultry sales. Use of pork has remained rather constant. Chicken consumption, including broilers, roasters, stewing hens, capons, etc., has grown spectacularly. Concealed is the rapid rise in broiler production and a decline in use of other types of chicken. Rice (1967) found broiler per capita consumption increased from 2 to 29 lb; uses of other types of chicken declined from 12 to 4 lb.

Fats and Oils

Since 1930, the total usage of fats and oils has remained relatively constant at 45-48 lb per capita. But there have been marked shifts in the percentage of the market supplied by major categories of products. In 1930, 12.7 lb of lard represented 28% of the food, fat, and oil utilization. In 1960, the 6.4 lb of lard represented only 13.5% of the total. Increases in other fats and oils made up for the declines in the use of lard (and butter). In 1930, animal fats supplied 65% of the total food fats, compared to 29% in 1970. Corn oil production increased from 158 million in 1940 to 370 million pounds in 1970. About 250 million pounds were used in margarine. This represents a per capita increase from 1.2 to 1.9 lb. Due to increases in other vegetable fats, corn oil actually represents a lower percentage of vegetable food fat now than it did in 1940. Soybean oil used in margarine and shortenings increased from 300 million pounds in 1940 to 1500 million pounds in 1970—more than a 5-fold increase. Much of this increase occurred in the 1950-1960 decade, according to Rice (1967).

Frozen Foods

Another dramatic development has been in frozen products. Frozen citrus juices, introduced in the late 1940's, increased from less than 1 lb per person in 1948 to about 35 lb in 1970. During the same period of time, the use of fresh and canned citrus products declined. In 1970, frozen citrus product utilization was at least equal to that of fresh citrus fruit. Frozen citrus juices now account for more than 10% of all fresh and processed fruit in the United States. Continuing changes in the food industry have been identified by Tressler *et al.* (1968) and Desrosier (1970) to be in convenience and institutional frozen meal and submeal product areas.

As we have tried to show, on a national average the amounts of foods available per capita at the retail level have not changed very much during several generations; but there have been significant changes in the foods supplying each category.

A change in nutrient supply has also occurred. There has been a gradual increase in fat, a small increase in protein, and a decrease in carbohydrate, with a net effect of a slight decrease in energy content of foods. Considering the period of time covered and the great changes which have occurred in the activities of consumers, these figures are relatively stable.

It must be noted that most of these data represent food disappearance at retail and do not necessarily mean actual food consumption. Such is the nature of available data. It is well known that part is not used because of spoilage, part is lost during preparation, and part is discarded as table waste. It is also clear that each of these represents a major area requiring attention. Obviously the consumer needs more efficient feeding systems.

SHIFTS IN MEAL COMPONENTS

There is no longer any question that meal components are also involved in these changing food habits. A particularly visible change is taking place at the breakfast meal. There has been a clear trend in this country toward skipping breakfast. Two reasons are given: lack of time and desire to lose weight.

Numerous surveys of American breakfasts have shown that a complete nutritionally adequate breakfast with all family members eating together is not particularly important to most families (Rice 1967; Diehl 1969; Hase and Riter 1970).

Fifty years ago, the word "breakfast" might have made one think of a family gathering leisurely in the morning at the breakfast table. The father and children would sit together and wait for mother to serve them cheerfully with some substantial meal.

Recent reports indicate that today, $^3/_4$ of all U.S. families do not eat breakfast together. Members of these families usually eat alone or with 1 or 2 other family members. Of the men, 10% don't eat breakfast at all. In 1 out of every 4 households, 1 child eats alone without parental supervision. Breakfast is no longer leisurely; the average person spends less than 20 min at the breakfast table. In a majority of households, even this short time period is interrupted by the telephone, lunch preparation, finishing homework, or household chores (Diehl 1969).

Breakfast has become a do-it-yourself meal.

Government reports reveal that fewer than 20% of all family members eat a breakfast which is considered nutritionally adequate or complete. Almost 60% of all children eat less than adequate (as defined by the mother) breakfasts and another 6% go off to school with no breakfast at all. Among teenagers, breakfast inadequacy increased to 70%; while for 1 out of 4 adults, coffee and toast is the standard menu (Rice 1967).

On weekends, 40% of the housewives serve the nutritionally complete breakfast which they believe their families should have. As a result, the housewife has had her role change from that of a "chef" to that of a purchasing agent whose primary function is to keep the pantry filled with an adequate inventory of supplies for various family members. The average family no longer keeps 1 or 2 favorite brands of cereal on hand, it keeps 5 or 6.

Each individual member of the family has now assumed responsibility for breakfast and selects the foods he wants or prefers depending on the amount of allotted effort and preparation time.

Some family members may be concerned with special health problems or with weight reduction and choose their menus accordingly. It is estimated that 25-35% of all housewives are frequently involved in dieting and most find breakfast the easiest meal to cut down or to cut out (Rice 1967).

Recent reports issued by the U.S. Dept. of Agr. (Anon. 1969) support the notions of modification in consumer eating habits.

For example, the traditional breakfast usually contained cereal-based products including breakfast cereals, biscuits, crackers, and bread. Some indications of their usage are presented in Tables 2.4, 2.5 and 2.6, which reveal actual changes in the breakfast cereal-based markets between 1962 and 1967.

TABLE 2.4

CHANGES IN CEREAL-BASED PRODUCTS EATEN FOR BREAKFAST (1962-1967)

Breakfast Foods	Households Serving (%)	Change from 1962 to 1967 (%)	Reason for Change
Yeast breads, buns rolls, coffee cakes	95.4	+1.3	Convenience
Doughnuts, toaster products, sweet rolls	38.3	+16.7	Convenience, sweet tooth, income
Crackers	5.8	+20.7	Convenience, diet
Ready-to-eat cereals	71.7	−2.8	Skip breakfast, choice of other foods
Hot cereals	31.6	−20.8	Convenience, choice of other foods
Grits, other grains	7.1	−16.7	Convenience, choice of other foods
Pancakes, waffles, fried cakes	43.0	−4.8	Convenience (time)

Source: U.S. Dept. Agr.

Such reports also reveal that teenagers appear to prefer foods that are not part of the traditional breakfast menu. In addition, they no longer eat only 3 meals a day, as we once did, but now eat 5 or 6 or more small meals, including convenience foods, instantized products, and snack foods as their general daily food supply.

TABLE 2.5

IMPORTANCE OF SNACKING FOODS (JULY-DEC. 1967)

	Total Servings at Snack Occasions (%)
Coffee	14
Soft drinks	52
Cookies	34
Ice cream	50
Juices	13
Cake & pie	18
Saltines and plain crackers	20
Potáto chips	36
Nuts	85
Pretzels and others (chips, curls, etc.)	62

Note: Volume distribution may be different, as more individuals may eat a serving at other meals.
Source: U.S. Dept. Agr.

TABLE 2.6

CHANGING FOOD SERVICE HABITS IN U.S. HOMES

	Total (%)	Homemade (%)
Cake	−8	−24
Cookies	−10	−18
Pies	−12	−16
	Canned (%)	
Soup	+6	−27
Soup–Ingredient use		
Mushroom	+12	
Tomato	+31	
	Frozen (%)	
Multiple course meals (packaged tray dinners)	+75	

Source: U.S. Dept. Agr.

CHANGES IN INDIVIDUAL EATING HABITS

What about trends in individual eating habits? While specific food components for diets have been changing, so, too, have consumers changed their eating habits. Historically, in this country foods have been eaten in meals spaced in the traditional 3-meals-a-day pattern. Other countries have 1 or 2 main meals accompanied by several lighter meals. Nutritionists tell us that so long as food intake is distributed throughout the day, the pattern does not apparently matter except that it should be effective.

There is a trend in this country toward the skipping of meals, particularly breakfast. Meal skipping is particularly bad when it leads to the ever more prevalent 1-large-meal-a-day diet. Rice (1967) reported that there is evidence that the large quantity of food consumed in a large single meal tends to be more fattening than the same food distributed throughout the day. (This is believed due to a buildup of the physiological system of the body required for converting carbohydrates and proteins to fat following ingestion of large quantities of nutrients.)

Surveys of diets of teenagers, reported by Rice, show that $^1/_3$ to $^1/_2$ of the girls and about $^1/_5$ of the boys do not eat breakfasts. Federal studies show that 15% of grade school children go to school without breakfasts, and only a few have an adequate breakfast. Among adults, the missing of breakfast or the consumption of a very minimal meal seems to be the prevalent habit.

This poses a challenge for the manufacturers of breakfast items, as will be discussed in a later chapter, particularly with conventional old-line products.

Five trends in U.S. food supplies are clearly becoming recognized: (1) The trend toward convenience foods of all types. (2) The trend further into ready-to-eat cooked foods, or heat-and-eat foods, including complete meals. (3) The trend of eating away from home—approaching a meal a day each. (4) The change in number of eating episodes per day. (5) The increased use of vending machines for distribution of foods.

While these developments do not necessarily alter national food consumption, they change the form in which food is desired. Overall, however, there is growing concern for the conservation of nutrients during storage of cooked foods for serving or vending. In addition, there are a multitude of snack, beverage, and dessert items easily accessible, each tending to favor diets that are overloaded with nutritionally "empty" calories.

The multiplicity of all types of foods, while affording excellent opportunity for the consumer to select according to purse or palate, also complicates menu planning. Each is a challenge, as well, to industry; an opportunity for some, enormous struggle for others.

MARKET CHANGES

Meanwhile, there have been remarkable changes in the methods of marketing foods. We are in the middle of a revolution in this area. The major change, of course, is that foods sit on the shelves in large markets, in great numbers, and people select what they wish. This means that the selling of foods has changed from a man who could sell something from behind the counter, to a food product which gets tried, and if it fills a need, is bought again. One of the hopes of the new day is for repeated purchases, which is a measure of success for a new product. At the same time, the consumer is making industry very uncomfortable, because it may have worked hard at establishing a workable routine in production and marketing, only to find a product no longer meets consumer needs.

PRODUCT LIFE CYCLES

As a result of the above and related matters, business enterprises talk more and more about the life expectancy and cycle of products (Fig. 2.4).

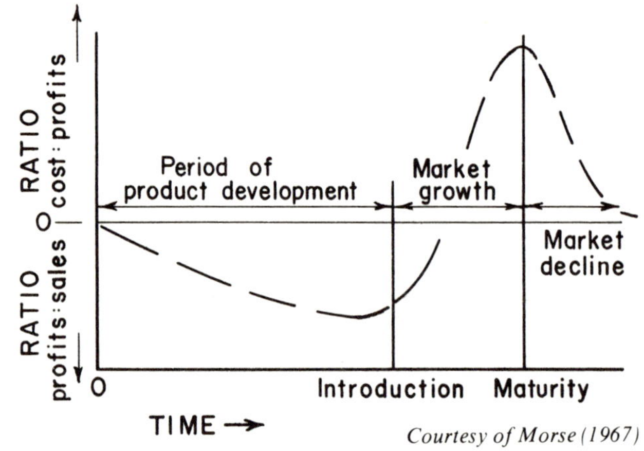

Courtesy of Morse (1967)

FIG. 2.4. LIFE CYCLE OF A NEW PRODUCT

This figure schematically represents the outflow and income of money versus the time to develop, produce, introduce, and sustain a new product on the market. All products have a life cycle. Some appear to be quite long; some too short. Nevertheless, they do occur.

Morse (1967) found that in the food industry, generally, the cost of development of a product during the laboratory stages is about 3-5% of the costs of bringing this product to the market. This is to the national market level. Compared with marketing costs, research expenditures are thus minor, generally. Investment increases as the product moves from the laboratory to the market. Morse estimated shifting from costs in the $30,000 range to the $2,000,000 range. This is one of the reasons why prudent effort is required in bringing a product from R&D to the market.

The maturity point in the life of today's product is getting shorter and shorter. Many feel that as time goes by, more and more conventional products will eventually disappear. Morse also believed that, more and more, products will be brought to the market with a predictable market life of "X" number of years. By this, it is suggested that in years, marketable life is going to start with, say, 5, and go down to 4, and then to 3 yr. Some feel it may become even less, more like experiences in the drug industry where the market life is quite short.

What does this mean to those in new product development? The industry will necessarily have to develop one product after another to keep in the marketplace, like waves hitting the beach. New product development—win, lose, or draw—appears to be here to stay.

ECOLOGICAL ADJUSTMENT OF BUSINESS

The business organization has many of the features of other living bodies. It may grow, it may join others, it may become more productive, and it may die. The environment in which it exists is necessarily changing, and these changes are critical to its existence.

Since the business environment is changing, change is becoming the way of life for the modern business enterprise. How it reacts to changing stimuli is evidence of its health and well being.

The concept of ecological adjustment of business to the changing environment, in a sense, is an extension of the marketing concept which in simple terms is consumer-oriented activity. When we bring into focus the environment in which the consumer lives and works and adjust our products and operations to meet changing demands, we are talking about ecological adjustment.

Neither ecological changes nor ecological adjustments take place overnight. The changes are necessarily gradual. Firms and organizations that adapt themselves, their products, and their search for new products stand the best chance to succeed in the long-run. If a company does not become a part of the solution, it becomes a part of the problem.

BIBLIOGRAPHY

Anon. 1965. U.S. Food Consumption: Sources of data and trends. U.S. Dept. Agr., Washington, D.C.

Anon. 1969. Food For Us All. Yearbook of Agriculture. U.S. Dept. Agr., Washington, D.C.

Desrosier, N. W. 1970. Technology of Food Preservation, 3rd. Edition. Avi Publishing Co., Westport, Conn.

Diehl, R. K. 1969. Market segmentation vs. positioning in product analysis. Food Prod. Develop. *4,* No. 2, 16-20, 64.

Gordon, T. J., and LeBleu, R. E. 1970. Employee benefits, 1970-1985. Harvard Business Rev. *48,* No. 1, 93-102.

Hase, P. F., and Riter, C. B. 1970. Building a technology into a business. Food Prod. Develop. *4,* No. 2, 16-20, 64.

Lonergan, B. J. F. 1957. Insight. Philosophical Library, New York.

Morse, R. E. 1967. The humility of product development. Food Prod. Develop. *1,* No. 2, 25-26, 55.

Rice, E. E. 1967. American dietary habits. Food Prod. Develop. *1,* No. 2, 21-24, 58.

Steiner, G. A. 1970. Rise of the corporate planner. Harvard Business Rev. *48,* No. 5, 133-139.

Tressler, D. K., Van Arsdel, W. B., and Copley, M. J. 1968. The Freezing

Preservation of Foods, 4th Edition, Vol. I, II, III, and IV. Avi Publishing Co., Westport, Conn.
Vancil, R. F. 1970. The accuracy of long-range planning. Harvard Business Rev. *48,* No. 5, 98-101.

CHAPTER 3

Market Forecasting

New products being the way of life for the modern corporation, how does the corporation explore the horizons available to it? How does it identify and forecast areas to enter? How does it make its decisions? These will be explored in this chapter.

Essentially, market forecasting can assist in the development of new products by supplying quantitative data and information on market trends. While there are many definitions of market forecasting, most have one thing in common: they can describe and provide marketing information essential to the branches of an industry.

Bouthilet and Weiss (1967) found three interlocking activities which could be subjected to forecasting: Raw material production, ingredient production, and consumer product production. It is the pattern of soy beans to oil to salad dressing, or wheat to flour to cookies.

Actually, the flow of events in the food industry is in the opposite order. For example, a company is organized to market consumer food products. It usually draws its supplies from a number of ingredient suppliers, who draw their raw materials from the basic agricultural sources. Let us then examine each in this order.

CONSUMER PRODUCT FIELD FORECASTING

Market forecasting is done at the first level by measuring consumer wants and needs. The consumer product industry is opportunist. Any single product on the supermarket shelf must compete with some 10,000 others to get into the shopper's grocery cart. Changes of less than 1% in product movement off shelves may mean a loss or gain of millions of dollars in gross sales. Constant market appraisal is necessary.

As will be discussed later, market researchers find out why, what, where and when consumers buy a certain food by asking. Next, they convert this information to forecasts and predictions of what, where, how, and why they will probably buy during the next year, etc. In order to predict effectively, this data must be combined with information on population and cultural trends. There are excellent sources available from the U.S. Bureau of Census and the U.S. Bureau of Labor Statistics. What results is a forecast of the consumer product field ahead.

INGREDIENT SUPPLY FIELD FORECASTING

Information required by this group is principally in the nature of basic forecasts. For example: What additives will be required? What processes will be used? What packaging systems will be used next year, etc.? What new ingredients will come into demand?

Since the ingredient supplier must establish the safety and advantages of his product, the supplier must plan in advance of the market.

For workable answers, Bouthilet and Weiss (1967) found that ingredient firms apply what economists call partial forecasting. They develop trends based, not on consumption, but on the knowledge that economies tend to become more and more industrialized.

Information sources to support partial forecasting are obtained by direct inquiries to the consumer product manufacturers, government statistics, and production data from pertinent trade journals. They find no single source of information adequate to provide needed data for forecasting.

RAW MATERIAL FIELD FORECASTING

Raw material suppliers are dependent on nature to provide crops and on the consumer product industries and/or ingredient suppliers for markets. As a result, basic raw material suppliers cannot respond quickly to changing demands.

The agricultural raw material suppliers obtain their basic statistics through the U.S. Dept. of Agr. and its Foreign Agricultural Services. In the United States, price support programs, acreage allotments, etc., regulate and control, to a large extent, production of many commodities. It is in the area of ingredient production that most companies need better agricultural marketing information.

As a result, many companies tend to generate their own ideas from a variety of sources. These include all the above, plus a staff which keeps abreast of changing situations in world markets as well.

FORECASTING WITH INPUT-OUTPUT ANALYSIS

Dash *et al.* (1967) reported that input-output analysis was not new, but that its use is rapidly expanding. The first officially compiled input-output Table of the United States economy became available about 15 yr ago. The most recent Tables provide information on more than 300 industries.

Sources of information for it include the U.S. Bureau of Census, U.S. Dept. of Agr., and the U.S. Bureau of Labor Statistics, plus information from consumer product companies, ingredient suppliers, and new material producers.

Leonteif (1966) described input-output analysis as a view of the flow of goods from one industry to another or from various groups within an industry. The data are usually presented in a device called a Transactions Table. It describes the flow of goods of one industry or group to another industry or group. Input-output analysis is believed by the experts in the field to provide a basis for studying the effect that changes in one sector of industry have or may have on any other sector of an industry.

The following example has been made available by Bouthilet and Weiss (1967). Their input-output diagram is shown in Table 3.1.

The left side of the Table lists the industries to be covered. Across the top of

TABLE 3.1

TRANSACTIONS TABLE (MILLION DOLLARS)

Inputs	Outputs				
	SIC 2022	SIC 2023	SIC 2026-2027	Final Demand	Gross Output
SIC 2022–cheese, natural and processed	17	1	12	521	551
SIC 2023–condensed and evaporated milk	4	1	48	635	688
SIC 2026-2027–fluid milk and other dairy products	381	363	2395	1448	4587
Other intermediate inputs	57	158	617		
Value added	92	165	1515		

Source: Bouthilet and Weiss (1967).

the Table each industry is again listed as column headings. For this example, they arbitrarily selected three interrelated industries: cheese, natural and processed; condensed and evaporated milk; and fluid milk and other dairy products. These they found listed in the Standard Industrial Classification Manual[1] as SIC numbers 2022, 2032, and 2026-2027, respectively.

As they explain, every cell in this grid simultaneously stands for either an input or an output. From the standpoint of its position in the vertical column, the number in a cell is an input to the industry coded at the top of the column. From its position in the horizontal row, the same number stands for an output of the industry coded at the left of the row. The total inputs and outputs of each industry in the columns and rows will always be identical. In their example in Table 3.1:

SIC 2022 sells $17 million within its own industry, $1 million to SIC 2023, $12 million to SIC 2026-2027, and $521 million to consumers for a total gross output worth $551 million.

SIC 2023 sells $4 million to SIC 2022, $1 million within its own industry, $48 million to SIC 2026-2027, and $635 million to consumers for a gross output worth $688 million.

SIC 2026-2027 sells $381 million to SIC 2022, $363 million to SIC 2023, $2395 within its own industry, and $1448 million to consumers for a gross output worth $4587 million.

They next calculated coefficients by dividing each cell in an industry column by the total output of that industry. From Table 3.1:

$$\$17 \text{ million}/\$551 \text{ million} = 0.0308$$

which is the input coefficient shown in the first cell of Table 3.2. The other

[1] Available from Office of the Bureau of the Budget, Executive Branch, Washington, D.C.

coefficients were calculated in a similar manner and became the direct coefficient matrix when assembled in Table 3.2.

TABLE 3.2

DIRECT COEFFICIENT TABLE

Industries Producing	Industries Purchasing SIC 2022	SIC 2023	SIC 2026-2027
SIC 2022	0.0308	0.0015	0.0026
SIC 2023	0.0073	0.0014	0.0105
SIC 2026-2027	0.6915	0.5276	0.5221

Source: Bouthilet and Weiss (1967).

This direct coefficient matrix revealed that a certain dollar's worth of material from one industry was required to achieve a certain dollar's worth of sales of another industry. In the case shown by Bouthilet and Weiss, \$3.08 worth of cheese products was required to achieve \$100 of cheese sales. Or put it another way, \$0.15 worth of inputs from the cheese industry was required to achieve condensed and evaporated milk industry sales worth \$100.

By inverting the direct coefficient matrix in Table 3.2, they developed the matrix shown in Table 3.3. This matrix gives us a measure of both direct and indirect effects, sometimes called a "multiplier." In the third cell of the first row of the Table, find 0.0057. This is to be compared to 0.0026 in the direct matrix shown in Table 3.2. This revealed to them that a total input of \$0.57 from the cheese industry found its way into sales of \$100 by the fluid milk industry. This compares with \$0.26 sold directly to the fluid milk industry by the cheese industry.

For the purpose of forecasting, Bouthilet and Weiss assumed that all of the sectors of the food economy increased at different rates. What would the pro-

TABLE 3.3

DIRECT AND INDIRECT REQUIREMENTS PER POUND OF FINAL DEMAND

	SIC 2022	SIC 2023	SIC 2026-2027
SIC 2022	1.0278	0.0045	0.0057
SIC 2023	0.0233	1.0053	0.0221
SIC 2026-2027	1.5129	1.1164	2.1087

Source: Bouthilet and Weiss (1967).

jected production and demand be for the products of these industries? In order to do this, they assumed a final demand for the examples under discussion, cheese (SIC 2022), canned milk (SIC 2023), and the fluid milk, etc. (SIC 2026-2027), at \$625, \$667, and \$1593 million, respectively, from available industry and government statistics. They also assumed that the inputs from each industry into the others would remain at the same ratio. The results of their analysis are shown in Table 3.4, the final demand for some future period. The projected interindustry transactions are shown in Table 3.5. They indicated that these are to be compared with the interindustry transactions of the original economic system (Table 3.1) to plan for the expanded industry requirements.

TABLE 3.4

PROJECTED TRANSACTIONS TABLE WITH CHANGES IN FINAL DEMAND

	Outputs				
Inputs	SIC 2022	SIC 2023	SIC 2026-2027	Projected Final Demand	Projected Gross Outputs
SIC 2022	20	1	13	625	659
SIC 2023	5	1	53	667	726
SIC 2026-2027	452	380	2636	1593	5061

Source: Bouthilet and Weiss (1967).

TABLE 3.5

ASSUMED CHANGES IN FINAL DEMAND

Industry	Original Final Demand	Projected Final Demand	Percent Change
SIC 2022	521	625	+20
SIC 2023	635	667	+5
SIC 2026-2027	1448	1593	+10

Source: Bouthilet and Weiss (1967).

While this presentation on input-output analysis has been abstracted for demonstration purposes, the value of such an approach is obvious. It can also be used to estimate the total impact of technological change. Bouthilet and Weiss report that 1 company is now using a 287 company matrix, and even this is not as detailed as they would like it to be. The size of the matrix desired would be a function of time, effort, cost, available information, and what results are needed.

TABLE 3.6

FACTOR AND SUBFACTOR RATINGS FOR A NEW PRODUCT

	Very Good	Good	Average	Poor	Very Poor
Marketability					
Relation to present distribution channels	Can reach major markets by distributing through present channels	Can reach major markets by distributing mostly through present channels, partly through new channels	Will have to distribute equally between new and present channels, in order to reach major markets	Will have to distribute mostly through new channels in order to reach major markets	Will have to distribute entirely through new channels in order to reach major markets
Relation to present product lines	Complements a present line which needs more products to fill it	Complements a present line that does not need, but can handle, another product	Can be fitted into a present line	Can be fitted into a present line but does not fit entirely	Does not fit in with any present product line
Quality/price relationship	Priced below all competing products of similar quality	Priced below most competing products of similar quality	Approximately the same price as competing products of similar quality	Priced above many competing products of similar quality	Priced above all competing products of similar quality.
Number of sizes and grades	Few staple sizes and grades	Several sizes and grades, but customers will be satisfied with few staples	Several sizes and grades, but can satisfy customer wants with small inventory of nonstaples	Several sizes and grades, each of which will have to be stocked in equal amounts	Many sizes and grades which will necessitate heavy inventories
Merchandisability	Has product characteristics over and above those of competing products that lend themselves to the kind of promotion, advertising, and display that the given company does best	Has promotable characteristics that will compare favorably with the characteristics of competing products	Has promotable characteristics that are equal to those of other products	Has a few characteristics that are promotable, but generally does not measure up to characteristics of competing products	Has no characterisics at all that are equal to competitors' or that lend themselves to imaginative promotion
Effects on sales of present products	Should aid in sales of present products	May help sales of present products; definitely will not be harmful to present sales	Should have no effect on present sales	May hinder present sales some; definitely will not aid present sales	Will reduce sales of presently profitable products
Durability					
Stability	Basic product which can always expect to have uses	Product which will have uses long enough to earn back initial investment, plus at least 10 yr of additional profits	Product which will have uses long enough to earn back initial investment, plus several (from 5 to 10) years of additional profits	Product which will have uses long enough to earn back initial investment, plus 1 to 5 yr of additional profits	Product which will probably be obsolete in near future
Breadth of market	A national market, a wide variety of consumers, and a potential foreign market	A national market and a wide variety of consumers	Either a national market or a wide variety of consumers	A regional market and a restricted variety of consumers	A specialized market in a small marketing area

Resistance to cyclical fluctuations	Will sell readily in inflation or depression	Effects of cyclical changes will be moderate, and will be felt after changes in economic outlook	Sales will rise and fall with the economy	Effects of cyclical changes will be heavy, and will be felt before changes in economic outlook	Cyclical changes will cause extreme fluctuations in demand
Resistance to seasonal fluctuations	Steady sales throughout the year	Steady sales—except under unusual circumstances	Seasonal fluctuations, but inventory and personnel problems can be absorbed	Heavy seasonal fluctuations that will cause considerable inventory and personnel problems	Severe seasonal fluctuations that will necessitate layoffs and heavy inventories
Exclusiveness of design	Can be protected by a patent with no loopholes	Can be patented, but the patent might be circumvented	Cannot be patented, but has certain salient characteristics that cannot be copied very well	Cannot be patented, and can be copied by larger, more knowledgeable companies	Cannot be patented, and can be copied by anyone
Productive ability					
Equipment necessary	Can be produced with equipment that is presently idle	Can be produced with present equipment, but production will have to be scheduled with other products	Can be produced largely with present equipment, but the company will have to purchase some additional equipment	Company will have to buy a good deal of new equipment, but some present equipment can be used	Company will have to buy all new equipment
Production knowledge and personnel necessary	Present knowledge and personnel will be able to produce new product	With very few minor exceptions, present knowledge and personnel will to able to produce new product	With some exceptions, present knowledge and personnel will be able to produce new product	A ratio of approximately 50-50 will prevail between the needs for new knowledge and personnel and for present knowledge and personnel	Mostly new knowledge and personnel are needed to produce the new product
Availability of raw materials	Company can purchase raw materials from its best supplier(s) exclusively	Company can purchase major portion of raw materials from its best supplier(s), and remainder from any one of a number of companies	Company can purchase approximately half of raw materials from its best supplier(s), and other half from any one of a number of companies	Company must purchase most of raw materials from any one of a number of companies other than its best supplier(s)	Company must purchase most or all of raw materials from a certain few companies other than its best supplier(s)
Growth potential					
Place in market	New type of product that will fill a need presently not being filled	Product that will substantially improve on products presently on the market	Product that will have certain new characteristics that will appeal to a substantial segment of the market	Product that will have minor improvements over products presently on the market	Product similar to those presently on the market and which adds nothing new
Expected competitive situation—value added	Very high value added so as to substantially restrict number of competitors	High enough value added so that, unless product is extremely well suited to other firms, they will not want to invest in additional facilities	High enough value added so that, unless other companies are as strong in market as this firm, it will not be profitable for them to compete	Lower value added so as to allow large, medium, and some smaller companies to compete	Very low value added so that all companies can profitably enter market
Expected availability of end users	Number of end users will increase substantially	Number of end users will increase moderately	Number of end users will increase slightly, if at all	Number of end users will decrease moderately	Number of end users will decrease substantially

Source: O'Meara (1961).

For further information, selected references are provided at the end of this chapter.

With the cost of computer times as low as it currently is and with computer programs available, input-output analysis seems destined to become a basic source of market forecasting information, and of growing importance in planning market development, diversification plans, checks on raw materials, intermediate supply and demand, and related matters, all of significance in new product development.

FORECASTING PRODUCT OPPORTUNITIES

Market forecasting, thus, can reveal trends and forecast market opportunities. Adding clarity and system to the potential product judgments involved was the focus of O'Meara (1961). He tried to cope with the selection of new products for development revealed by forecasting, with techniques and concepts developed in other fields of management. The procedure is distinguished by several main features:

(1) By utilizing simple probability and weighting techniques, the procedure reduces the overall new product problem to a series of simpler ones which are more easily and objectively solved.

(2) The solutions to these smaller problems are then combined into three index numbers that represent an overall rating for the potential of the new product.

(3) The first index number is a rating for intangible factors (not easily quantifiable), such as marketability, growth potential, productive ability, and durability. The second index number indicates the product's short-term profitability or payback potential. The third number indicates the product's long-run profit potential.

Intangible Factors

Of the three index numbers to be derived, the first is for intangible marketing factors. To illustrate the procedure, O'Meara used a hypothetical case involving Company Y and a new product, Product X.

Assume that Company Y chooses four major intangible factors, divides each of them into its component subfactors, and defines them as in Table 3.6. Another company with different types of products might break the problem down differently, but essentially the same procedure should be used.

The next step, according to O'Meara, is to study the four major factors and weight them so as to indicate the evaluator's judgment of their importance in relation to each other. These relative weights for Company Y are indicated in Table 3.7, Column 2. Similarly, each of the subfactors that comprise the four major factors was weighted. (For the sake of brevity, O'Meara used only the subfactors of marketability to illustrate this procedure.) At the same time, numerical values are assigned to the definitions transferred from Table 3.6.

TABLE 3.7
SUMMARY SHEET OF INTANGIBLES

Proposed product: Product X *Evaluated by:.* John Smith

1 Factor	2 Factor Weight	3 Assigned Factor Value	4 Final Factor Evaluation
Marketability	0.4	71.4	28.6
Durability	0.3	68.6	20.6
Productive ability	0.1	91.6	9.2
Growth potential	0.2	69.2	13.8
Final intangible factor index number			72.2

Source: O'Meara (1961).

Company Y has thus constructed a framework that enables it to evaluate the overall problem of new products more objectively by breaking it into its component parts, the factors and the subfactors. Weights have been assigned to these components to indicate their relative importance. At different times, the relative importance of these components may vary; if so, the weights would then have to be changed accordingly, O'Meara pointed out.

Estimated Probability

The next step in O'Meara's procedure is to study more thoroughly the description of each of the subfactors, and to enter in Table 3.8, Columns 3 through 7, the estimated probabilities (EP) that the product in question will equal the description of "very good," "good," "average," "poor," or "very poor." This is the first instance in which the company actually evaluates the proposed product. The purpose, he points out, is to permit an evaluator to express, in numbers, exactly what he thinks the product's possibilities may be. For example:

If one believes there is at least a 50-50 chance that the merchandisability characteristics of Product X will meet the definition of "very good," that there is less chance that it will meet the definition of "good," and that there is an even smaller chance that it will meet the definition of "average," he also believes that there is no chance at all that the definitions of "poor" and "very poor" will describe the merchandisability of the product.

He therefore enters probabilities of 0.5 for "very good," 0.4 for "good," and 0.1 for "average," allowing no probability for "poor" or "very poor."

This method of precisely stating one's best judgment, according to O'Meara, will result in a more efficient evaluation than would be possible using a less systematic procedure. This will be discussed further below.

TABLE 3.8

EXAMPLE OF THE USE OF AN EVALUATION SHEET

Proposed product: Product X *Evaluated by:* John Smith

Factor: Marketability

1	2	3		4		5		6		7		8	9
	Sub- factor	Very Good (10)		Good (8)		Average (6)		Poor (4)		Very Poor (2)		Total	Subfactor Evaluation
Subfactor	Weight	EP[1]	EV[1]	EP	EV	EP	EV	EP	EV	EP	EV	EV	Col 2 X Col 8
Relative to present distribution channels	1.0	0.1	1.0	0.2	1.6	0.5	3.0	0.2	0.8	–	–	6.4	6.4
Relative to present product lines	1.0	0.1	1.0	0.2	1.6	0.4	2.4	0.2	0.8	0.1	0.2	6.0	6.0
Quality/price relationship	3.0	0.3	3.0	0.4	3.2	0.2	1.2	0.1	0.4	–	–	7.8	23.4
Number of sizes and grades	1.0	0.1	1.0	0.2	1.6	0.5	3.0	0.2	0.8	–	–	6.4	6.4
Merchandisability	2.0	0.5	5.0	0.4	3.2	0.1	0.6	–	–	–	–	8.8	17.6
Effects on sale of present products	2.0	–	–	0.2	1.6	0.5	3.0	0.3	1.2	–	–	5.8	11.6
	10.0											Total factor value	71.4

Source: O'Meara (1961).

[1] EP = estimated probability as judged by managment; EV = expected value computed by multiplying the rating's numerical value by the estimated probability.

Expected Value

After assigning probabilities to the subfactors, multiply each probability figure by the numerical value attached to the rating. The results are placed in Columns 3 through 7 in the divisions for expected value (EV). Again, O'Meara uses merchandisability as an example:

Since it has been estimated that there is an 0.5 chance that Product X's merchandisability will meet the definition of "very good," and the numerical value is 10.0, the expected value is 5.0. This figure, then, is entered in the EV column. He also estimates that a 0.4 chance exists that the product's merchandisability characteristics will equal the definition of "good," and that a 0.1 chance of equaling the definition of "average" exists. Multiplying these probabilities by the numerical values attached to their respective definitions, he obtains the expected values (EV) that are shown in Columns 4 and 5 for merchandisability.

Up to this point O'Meara focuses on dividing the problem into manageable segments and allowing objective decisions to be made. The next step in his system is to add the individual EV for each subfactor and enter the total in Column 8. This step begins the process of combining many decisions and judgments that have been made, with the ultimate aim of reducing them to one major index number. O'Meara suggests that the numbers in Column 8 should represent the best numerical estimate that can be obtained of the degree to which the characteristics of the proposed product meet the desired characteristics listed as subfactors in Table 3.6.

The next step, using the O'Meara technique, is to multiply the total EV by the weights that had previously been assigned to their respective subfactors (Column 2). These results are then entered into Column 9. The total at the bottom of Column 9 represents the value of the product's marketability as a whole. This major factor value, as it might be called, is entered in Column 3 of Table 3.7.

He follows the same procedure to get the values for the other three major factors -durability, productive ability, and growth potential. Assume the values shown in Table 3.7 are thus generated.

Then, O'Meara multiplies these values by the weights given earlier to the major factors. If, for example, marketability were considered more important than productive ability, one would want to carry that difference into the final evaluation. Accordingly, O'Meara finds that each figure in Column 4 combines the series of judgments about how well the product will measure up (Column 3) with judgments about how important the measure is to the company in comparison with other measures (Column 2).

The total at the bottom of Column 4 in O'Meara's method is the index number that represents the final evaluation of how well Product X fulfills the intangible requirements for a new product and represents decisions on many problems, each of which has been weighted in the light of its importance to the total problem.

Short-Run Profitability

The second index number needed is one that indicates the product's short-run profitability, which O'Meara refers to as the "payback index." This index is obtained by establishing the product's probable short-run profit, and dividing this estimate by the development cost. The O'Meara formula is shown at the top of Table 3.9. In effect, payback gives the time required for short-run profits to equal the development expenditures.

TABLE 3.9

PAYBACK INDEX

Proposed product: Product X *Evaluated by:* John Smith

Formula: $\frac{A \times B \times (C - D) \times E}{-F + G + H + I}$

Factors	
Probability of commercial success	70%
Estimate of sales–units per year (avg)	$250,000.00
Estimate of selling price per unit	$5.00
Estimate of cost per unit	$4.00
Competitive grace period	3 yr
Additional working capital needs	$10,000.00
Market development cost estimate	$10,000.00
Additional capital expenditures	–
Production development cost	$30,000.00

$$\frac{0.7 \times 250{,}000 \times (5 - 4) \times 3.0}{10.000 + 10{,}000 + 0 + 30{,}000} = 10.5 = \text{payback index}$$

Source: O'Meara (1961).

The general purpose of the formula is revealed in the factors used for the numerator and the denominator. They are: (1) Probability of commercial success—estimated by an experienced group. (2) Estimate of average sales unit per year—estimated by marketing manager. (3) Estimate of selling price per unit—estimated by marketing manager. (4) Estimate of cost per unit—estimated by production manager. (5) Competitive grace period—estimated by development group. (6) Additional working capital needs—estimated by development group. (7) Market development cost estimate—estimated by marketing manager. (8) Additional capital expenditures—estimated by production manager. (9) Production development cost—estimated by production manager.

The probability of commercial success and the competitive grace period need further description.

The figure for the probability of commercial success places a limit on confidence in the sales projection. Since the sales figure is to be estimated, the probability figure permits a further judgment to be made by the development group, hence usually resulting in a more carefully scrutinized index number, according to O'Meara.

The competitive grace period is defined by O'Meara as the period of time during which no major, adverse changes on the product's market are expected. Because of market uncertainties, the grace period should be limited to a maximum of 4 yr, hence setting a limit on the length of the payback period to be considered.

Assume the values shown in Table 3.8 and use them in the payback formula. This reveals an index of 10.5 for Product X. If similar indices for other new products under consideration are generated, we will have another helpful yardstick to use in making comparisons and in deciding which items should be pursued.

Long-Run Profit Margins

The final index number used in the O'Meara method is one that represents long-run profit margins. (Long-run to mean longer than 5 yr, and profit margins to mean steady margins that can be expected when the product assumes its normal place in the market.) Since this estimate deals with occurrences far in the future, O'Meara found it not as reliable as some of the estimates for intangible factors and short-run profitability, as discussed above. While he recognized that less emphasis can be placed on this number, it does indicate the long-range potential of the product, and should be used to complete the overall evaluation.

O'Meara illustrated how the index worked out in Table 3.10:

TABLE 3.10

LONG-RANGE PROFITABILITY TABLE

Product: Product X *Evaluated by:* John Smith
Assume: 100,000 Sales

1 Profit Margin (%)	2 Probability	3 Conditional Profit ($)	4 Expected Profit ($)
10	0.1	10,000	1000
15	0.1	15,000	1500
20	0.3	20,000	6000
25	0.3	25,000	7500
30	0.2	30,000	6000
	Total expected profit		$22,000
	Total expected profit as a percentage of sales		$\frac{\$22{,}000}{\$100{,}000} = 22\%$

Source: O'Meara (1961).

Column 1 lists the most logical range of profit percentages that can be expected. In Column 2, one estimates the probability that these different profit percentages will be attained.

The conditional profits are entered in Column 3. For a 10% profit margin on a $100,000 sales estimate the conditional profit would be $10,000; for a 15% profit margin, $15,000; and so on.

Column 4 is the product of the numbers in Columns 2 and 3. For example, there is a 1-in-10 probability that a $10,000 profit will be made, and therefore the expected profit is $1000.

The total of Column 4 is used as a numerator, and the assumed $100,000 sales is used as a denominator, which fraction is then converted to a percentage figure. This gave O'Meara an estimate of the product's long-range profitability.

Hence, he developed a framework with which one can express in figures what he thinks to be the product's potential. Next, he combines the foregoing figures into one long-range profitability index, such as shown in Table 3.10.

The system developed by O'Meara is flexible both in structure and in application. At any time, as he has pointed out, the framework can be altered either to include factors that have increased in importance or to eliminate those which have decreased in importance. For example, if the choice of factors and subfactors in Table 3.6 becomes outmoded, the list can easily be changed to suit current conditions. Also, if the ranges of factors in Table 3.6 should turn out not to be applicable to a proposed product, it could be changed to meet the situation.

In any event, the O'Meara method contains many important and useful insights in that they can be of great value in taking market forecasts one step further, in a sensitive and effective manner.

LONG-RANGE PLANNING

In association with the above, it is possible next to identify the market interests of divisions of a corporation, using the information generated by the market forecasts. These general interests can next be subjected to long-range planning techniques to establish which of the areas identified are the most appropriate to initiate and why. The techniques have been used widely in locating areas for major corporate thrusts.

Enlisting Broad Support

In planning, it is imperative to have broad participation. Many long-range planning efforts had been ineffective, principally because of failure to enlist broad participation and provoke wide discussion in a company. To avoid this, according to McGlauchlin (1968), three principal sources are needed to provide necessary inputs: (1) The divisional corporate groups that eventually make use of the results. (2) The research scientists who are in touch with new scientific trends. (3) The top corporate financial, legal, and administrative staffs. To initiate the program, each participant must be persuaded to recognize the explicit invitation to make his opinions known.

Each is asked to:

(1) List 2 or 3 areas of the greatest importance to his division. This list includes only those areas so vital to his operation that he would desire to move in this direction.

(2) Indicate the relative importance of each need and state in several paragraphs how and why this area could improve corporate profit.

(3) Include a brief discussion of alternative solutions. This information should contain a report on what has been tried, and the merits and shortcomings of the obvious modes of attack.

(4) Delegate fairly broadly the task of making recommendations, but make sure that the requests represent a joint effort with marketing, and that the requests have also been reviewed and approved by the power structure within the division of the corporation.

Each division supplies recommendations from several different individuals. The director of marketing in one division might ask for research on some problem, and, a short time later, the director of production in the same division would label a different area of research as the one most urgently needed. The requested work tends to stress those needs related to current production, engineering, or sales problems, according to McGlauchlin (1968).

(5) Select only those items that represent persistent needs.

(6) Make sure that the topics cited are suitable for effort.

Research Laboratory Recommendations

Certain broad principles in gathering recommendations from scientists are needed. First, they must predict scientific and technological trends. Second, they must indicate what knowledge they hope to gain and how this increased understanding is related to present or future business. In other words, clear objectives for the work are essential (McGlauchlin 1968).

They must estimate the manpower needed and make up a schedule for expected intermediate results, and draw an approximate curve showing the probability of attaining a given research objective. This should show the estimated minimum time to reach the specified goal, the most probable time needed, and an approximate maximum time needed if unforeseen difficulties arise. It also indicates the manpower needed at various stages of the project, stated in terms of scientific teams. For this purpose a team consists of a senior researcher, a research assistant, and one or more technicians. The curve method encourages the team to make a series of judgments, McGlauchlin found.

Corporate Management's Part

Corporate management provides a suggested rate of growth for the overall expenditures during the planning period—that is, a trend line for budgets.

Corporate management also proposes concentrating on a few technological areas that appear attractive as a result of the exploratory work done in prior years.

Combining the Inputs

Having received the inputs from the groups, one next determines the overlap among the needs and makes a company-wide ranking of the proposed work, as outlined by McGlauchlin (1968).

It is to be expected that a substantial number of divisions view their needs in relation to current products. While one division's products and services often differ considerably from those of others, he found, it becomes apparent that one program can usually serve several needs, since the knowledge generated hopefully would have broad application.

The net effect is a shortening of the list by using one program to meet several needs.

In making a company-wide ranking of the recommended programs, McGlauchlin next places primary emphasis on two factors: (1) The corporation's market penetration at present and in the future. (2) Whether the market area in question would be dominated by price competition or by performance factors.

Market penetration is estimated by obtaining data and information from sources described in market forecasting for various classes of products or services, and comparing these with the company's own data. The questions regarding performance competition or price competition are resolved by considering the nature of the customer's needs and the products available from competitors. Marketing managers supply this information.

Price versus performance can be defined as follows:

Performance Competition.—The objective is a performance advantage of the proposed product compared with competing products in the market.

Price Competition.—The objective is a slight economic advantage over several close competitors.

Regarding the first factor, if the company commanded a very large part of a specific market, the potential gain in market penetration or market share would be correspondingly small. Regarding the second factor, payoffs are largest when the results lead to improved performance or provide a service urgently needed but previously unavailable.

McGlauchlin combines the inputs by evaluating the proposals in terms of their suitability for a company. This phase eliminates those programs that require (1) large and expensive facilities not available, (2) a good understanding of the eventual user of the product, or (3) an intimate knowledge of high-volume production techniques.

Rating Current Efforts

Before we proceed to formulate a long-range plan, it is necessary to evaluate current programs to develop a systematic yardstick for evaluation. Ideally, a quantitative method, such as a calculation of the return on funds invested on past projects would be preferred. New product development activities lend themselves very readily to quantitative assessments. Inasmuch as the amount of money a company spends on research is substantial, research management can hardly be both judge and jury of their own case, no matter how objective they try to be.

McGlauchlin suggests the common effective way of evaluating things is by accepting the value judgment of the "buyers" of research results, that is, the company divisions.

Operating Division Evaluation

Next, he submits to the profit generating division managers a description of each of the currently proposed programs—their purposes, objectives, and possible applications. Then he asks for their evaluations, using the information he has submitted and any other knowledge gathered about these programs.

Each operating department might be asked to rate proposed programs, using two associated numbers. One, called "Combined relative value," *Rcv*, is described by McGlauchlin as follows:

$$Rcv = A_i S_i$$

Here A_i represents each division's rating of the applicability, rating how much of a program was or would be useful and how important that fraction was or is to the division. S_i is the size of the division measured in annual sales. The equation emphasizes the larger divisions' opinions and assigns less influence to the smaller operating parts of the company.

The second number needed (A_i) rates a program without any weighting related to a division's size.

Comparing these two values, *Rcv* and A_i, for any program with those of any other program provides a relative evaluation of all. Taking two programs, about equal in weighted values, the endorsement of one by large corporate components results in a far higher combined relative value for one program than for another which gains support from relatively small company units.

McGlauchlin reported that an individual rating might also be evaluated according to that division's anticipated growth. After consideration, he rejected this contention on the ground that it violated the accounting principle of conservatism. If the anticipated growth did not occur, the group would have been permitted an undue influence on the company's affairs; if the anticipated growth was achieved, the division's influence would keep pace.

Another method of evaluation is to project returns on investments. This area will be discussed in detail in Chap. 10.

Company Overview

By comparing the relative ratings of programs for which data exist on both relative value and potential profitability due to previous efforts, the potential value of a new program can be estimated. Placing "new" proposals on a scale with "existing" previous points serves as a guide in helping corporate management determine which potential new program should receive emphasis in terms of achieving corporate goals.

Thus, a company can generate the information needed to forecast long-range programs. Also, as a result, company management has a list from divisional managers of the areas where they need help; from research it has a list of investigations they would most like to make, from the divisions, an evaluation of the current research programs; from management itself comes guidelines on expenditures for the 5-yr planning period and the emphasis which should be placed on promising technological areas revealed by forecasts, and plans set accordingly. The reason is that if there is high expectation of a project's meeting its objectives, it is assumed that this is true for other companies also. Thus, the commercial benefits would probably be small because several companies would compete in offering products based on similar technical advances.

This notion was expressed by McGlauchlin in a formula in which the probability of success, P_S, times the value of the result, V_r, is equal to a constant:

$$P_S V_r = K$$

The relationship says that as P_S gets larger, V_r gets smaller, since the product remains the same.

Conversely, if the market area proposed requires a large advance in the state of technological art, the probability of technical success is small. The corollary is: if the objectives are achieved in spite of great obstacles, the company that sponsored the area activity could gain a considerable lead time in which to exploit its success. Other companies probably will not be able to duplicate the result quickly. Evaluation of all inputs therefore would stress those area projects which promise large strides forward, provided that they can be related to a clear and substantial market (i.e., if K is large).

At the same time, corporation managers must indicate the relative emphasis they place on growth by acquisition and growth through exploitation of research, including the potential effect of legislation and governmental policy. Financial and legal advisers similarly offer special competence in predicting the relative profitability of various types of business endeavors. In the end, the corporation leader must assess the rates of growth expected from the existing profit centers, and blend the whole to yield the maximum returns on investments, and a plan for the future growth of the enterprise.

To recapitulate briefly, we have explored market forecasting and long-range planning to achieve corporate growth. Given the area a corporation wishes to expand into, the next step is to generate ideas and test their value in terms of meeting the consumer market needs which have been forecasted, and corporate requirements for growth which are planned. How this is accomplished will be developed in the next chapter.

BIBLIOGRAPHY

Aguilar, F. J. 1967. Scanning the Business Environment. The Macmillan Co., New York.

Anon. 1968. Standard Industrial Classification Manual. Office of the Bureau of the Budget, Washington, D.C.

Argenti, J. 1969. Corporate Planning. Dow Jones-Irwin, Homewood, Ill.

Bouthilet, R. J., and Weiss, H. K. 1967. Market forecasting for new product development. Food Prod. Develop *1*, No. 2, 16-20, 48.

Cox, D. F., and Good, R. E. 1967. How to build a marketing information system. Harvard Business Rev. *45*, No. 3, 145-154.

Dash, J. K., Fuerst, J. M., and Rabitsh, E. K. 1967. Input-output as a planning tool at Celanese. Chem. Marketing Econ. Symposium, Am. Chem. Soc., Chicago, Ill. Sept., 1967.

Ferrell, R. W. 1964. Customer oriented planning. Am. Management Assoc., New York.

Goldman, M. R., and Marimont, M. L. 1964. The inter-industry structure in the U.S. Survey of Current Business. U.S. Dept. Com., Washington, D.C. Nov.

Katona, G. 1964. The Mass Consumption Society. McGraw-Hill Book Co., New York.

Leonteif, A. 1966. Quantitative input-output relations in the economic systems of the U.S. Rev. Econ. Statist. *18*, No. 8, 105-125.

Levitt, T. 1965. Exploit the product life cycle. Harvard Business Rev. *43*, No. 6, 81-94.

McGlauchlin, L. D. 1968. Long range technical planning. Harvard Business Rev. *46*, No. 4, 54-63.

Miernyk, W. H. 1966. Elements of Input-Output Analysis. Random House, New York.

O'Meara, W. H. 1961. Selecting profitable products. Harvard Business Rev. *39*, No. 1, 95-102.

Reynolds, W. H. 1965. More sense about market segmentation. Harvard Business Rev. *43*, No. 5, 107-114.

Steiner, G. A., and Cannon, W. M. 1966. Multinational Corporate Planning. The Macmillan Co., New York.

CHAPTER 4

The New Product Idea

We have reached the point where we must generate ideas to meet the targets identified by market analysis and forecasting, in one form or another. Generating ideas is straightforward, for it is an integral part of the nature of man. Generating ideas that will yield a successful new product, fulfilling broadly-based consumer needs, and at the same time meeting corporate growth requirements, is quite another matter. Realizing full well that there are probably as many specific systems being used as there are groups working at idea generation, we will try to develop an overall insight into what is presently being done to accomplish the task. There is a recurrent scheme, and to it we address ourselves.

IDEA GENERATION

Crockett (1969) identified the following general sequence of events in developing new product concepts. Essentially, the search involves identifying opportunity areas and the problems and needs within these areas, finding potential solutions to these problems and needs in the form of new product ideas, and their evaluation prior to being placed for work into the research laboratory. Setting aside the evaluation of ideas until the next chapter, let us explore the other elements in the sequence and see how they work. (See Table 4.1.)

TABLE 4.1

THE FIVE–STAGE FUNNEL LEADING TO R&D

Stage No. 1 Opportunity analysis	Identify industries, product classes, and consumer use situations
	↓
Stage No. 2 Problem areas	Identify problem areas in selected opportunity areas
	↓
Stage No. 3 Idea generation	Identify solutions to problem areas plus input of random ideas
	↓
Stage No. 4 Idea evaluation	Disciplined system of judging solutions (ideas)
	↓
Stage No. 5 Concept testing	Consumer reaction to ideas
	↓
	Input to R&D

Source: Crockett (1969).

Good ideas are ideas which are relevant to a company's needs, ideas of real commercial value. Crockett reports that the task of trying to sort out and evaluate the good from the bad is almost unmanageable. To enter laboratory development with most of these ideas, without proper evaluation, is a nearly certain road to disaster. The odds are something near 100 : 1 against a winner.

The "idea search" function must be removed from the random category to a disciplined basis with more favorable odds for success. The odds improve, essentially, when new product ideas are passed through some form of intellectual filter.

OPPORTUNITY ANALYSIS

This involves the identification of industries and product classes and consumer use situations that are most relevant to the company's needs.

The need at this point is to sort quickly through the almost limitless choices for those few which warrant further study in depth. This can be accomplished in many ways, according to Crockett:

(1) Make an industry/product class analysis of all secondary data sources, utilizing some preestablished criteria as guidelines (Table 4.2), as explored in the last chapter.

TABLE 4.2

CRITERIA FOR INDUSTRY/PRODUCT CLASS ANALYSIS

Criteria	Source
Size of industry	NNPS[1]/food topics
Its trend and future outlook	Food topics/trade sources
Its composition	NNPS/wholesale books
The product mix	NNPS/wholesale books
New product activity	Reports, advertising agencies
Sales methods	Sales department
Channels of distribution	Sales, trade sources

1 Nielsen New Product Source.
Source: Crockett (1969).

(2) Study other external forces which affect consumers at present or are expected to do so in the future. Exterior forces affecting consumers (Crockett 1969) are: the applied industry, the food distribution system, science and technology, supply/demand of basic commodities, and shorter workweek and more leisure time.

(3) Study family or consumer patterns, food uses and attitude information leading to their unfulfilled needs.

After screening and sorting out a large number of product classes and consumer situations, experience reveals there usually remain a few areas in which to concentrate efforts to identify the new product opportunities of greatest value.

LOCATING PRODUCT OPPORTUNITIES

There are a number of methods being used to identify new product opportunities generally composed of two main elements: what the market looks like to customers and what the product offered means to them.

The Traditional Method

The traditional approach according to Lavidge (1968) might be to see the food market in segments and, knowing production capabilities, develop ideas to fill a segment of interest to a company. A list of such segments would consist of

TABLE 4.3

TRADITIONAL MARKET SEGMENTS

Perishables
Baked goods
Dairy products
Frozen foods
Fresh and cured meat, fish, and poultry
Produce
Dry Groceries
Baking needs
Beverages, alcoholic
Beverages, prepared
Candy and chewing gum
Canned foods
Cereals, flour, macaroni
Condiments, dressings, spreads, relishes
Desserts
Dried foods
Fats and shortenings
Jams, jellies, preserves
Sweeteners and flavors
All other foods

Source: Lavidge (1968).

something like that shown in Table 4.3

Such lists tend to be oriented to products, distribution systems, or retail markets and, most believe, not specifically oriented to uncovering consumer needs.

The Trigger System

Trends can also provide suggestions to help come up with a new product idea. Some are trends in our population, trends in our habits, trends in discretionary spending of time and/or money.

If we are seeking new product ideas, it is usually worthwhile to organize the information we have available.

There are several sources of useful information: product users, product distributors, neighbors, family, government agencies, business associations, and industry suppliers. Each can stimulate a new product concept. The accompanying Table 4.4 lists some of these major triggers to help generate new product concepts. This list was developed by Lavidge (1968) and found to be useful in organizing relevant information. As he noted, if products are to be in tune with markets, and markets are changing, it is inevitable that we must offer products that are also changed. While it may lead to the modification or even elimination of some existing products, it may also generate opportunities for the addition of profitable new products.

TABLE 4.4

PHASE 1: CONCEPT GENERATION

	Information Status	
Triggers[1]	Available	Needed
Potential customers		
Unsatisfied wants		
Unrecognized needs		
Trends		
Technical developments		
New products		
New processes		
Basic research developments		
Packaging developments		
Functional		
Distribution		
Marketing methods		
Graphic		
Personal selling		
Promotion		
Pricing		

Source: Lavidge (1968).

1 One shortcut to organizing the existing or available data for use in a new product concept groups the source triggers into common associations. Regrouping or associating these in other ways can also stimulate new avenues.

Look for voids in the opportunity areas retained for further review. Competitors may have overlooked certain aspects. After reviewing the accumulated data uncovered in published literature and related sources, then sessions are scheduled with consumer panels of the appropriate makeup. Together, these will assist in establishing parameters on unfilled consumer needs, hence, opportunities.

The Meal Method

An example has been presented by Diehl (1970), and is abstracted below.

Diehl chose to ignore the traditional approach and created a market framework that hypothesized the way consumers view food products. He called this market framework, Meal-Market. Each of the daily meals was a focal point of consumer behavior and, hence, a potential market.

Meal-Market divided food products into four segments housewives use as they shop for and use food: breakfast, lunch, dinner, and supplementary (snacks, etc.).

The method is simple because it reflects the consumer's desire to simplify her purchasing decisions and patterns. For example, consumers might tend to nibble crackers at 3:30 in the afternoon because they want a snack, and probably not because crackers are another form of baked goods.

The Diehl method provides a productive way of looking at existing quantitative data. Information areas used in the Meal-Market breakfast segment are reported by him and are shown in matrix form in Table 4.5. The information comes from published data. He calculated the column headed "amount spent for breakfast consumption" by rating each product placed in the breakfast category in terms of the estimated consumption of each item at breakfast. For example, his rating terms were: virtually all (cold cereal), mostly (bacon), some (butter), and little (meat). Next, he applied a standard percentage for each term against total dollars spent in grocery stores.

Similar data were developed for the lunch, dinner, and supplementary use categories. As might be expected, this led to insights into product dynamics, and this became the key to productive new product idea generation.

Consumer Dynamics

In the example below, Diehl (1970) chose breakfast as a typical meal, insofar as consumers were concerned.

Breakfasts are eaten in three main locations: kitchen (80%), dining rooms (15%), and other rooms (5%). To $^1/_2$ of the housewives breakfast was not a particularly satisfying period. A complete breakfast was felt to consist of fruit or juice, cereal or potato product, and eggs, meat, or fish. He found that teenagers felt everyone should have a good breakfast; but $^1/_2$ of the older girls and $^1/_4$ of the older boys interviewed said they had either no breakfast or one without solid food.

Many women were not convinced that they, themselves, need three regular meals a day; consequently, they do not serve breakfast with great conviction. Women with children ate better breakfasts than those without; women who ate with their children had better breakfasts than those who didn't.

Based on such findings, he concluded that regardless of the product being considered the following were some key attributes of consumer breakfasts:

TABLE 4.5

MATRIX FOR EVALUATING TRADITIONAL BREAKFAST MARKET DATA WITHIN THE MEAL-MARKET FRAMEWORK

Products	Amount Spent for Breakfast Consumption	% Change, Total Grocery Store Sales 196X-196Y	% Change, Total Grocery Store Sales 196Y-196Z	% of Grocery Store Sales to Total Consumption	Total Spent in Grocery Stores for All Meals	% of Total Grocery Store Sales
Cold solids	(Million $)				(Million $)	
Cereals	466.3	+8.7	+7.0	95	548.6	0.91
Doughnuts						
Table syrup						
Jams, jellies, preserves						
Fresh fruit						
Sugar						
Butter						
Cakes and pastry						
Margarine						
Salt						
Pepper						
Canned fruits						
Frozen fruits						
Total						
Hot solids						
Cereals	70.9	+6.3	+1.8	90	83.4	0.14
Infant cereals						
Pancake flour						
Bacon, ham sausage, other						
Shell eggs						
Corn meal, grits						
White bread						
Canned baby foods						
Other bread						
Frozen meat						
Total						
Total solids						
Cold liquids						
Frozen fruit juices	285.8	+1.8	−3.0	86	336.2	0.56
Fresh milk						
Canned juices						
Fresh cream						
Dry milk powders						
Canned milk						
Total						
Hot liquids						
Prepared cocoa	15.6	+3.2	+3.2	61	22.3	0.04
Cocoa						
Coffee, ground						
Instant coffee						
Tea						
Total						
Total liquids						
Total spent for breakfast consumption					Total spent in grocery stores	

Source: Diehl (1970).

Convenience.—Inconvenience of preparation was a major obstacle in the housewife's serving a complete breakfast (see Table 4.6).

Taste.—Preferences were expressed for traditional breakfast forms and flavors.

Nutrition.—The majority of women seem to know what constitutes a nutritionally adequate breakfast even though they generally tend not to serve one.

Essentially, he had found that the ideal breakfast must deliver convenience, taste, and nutrition. Thus, the first set of product dynamics was established. Further work showed a strong appeal for breakfast products based on physical form.

TABLE 4.6

NEW PRODUCT RATING SYSTEM

1. Cost value score: $\frac{1}{\text{cost (¢) per serving}} \times 100$	= ______
2. Nutritional score[1]: $\frac{1}{\text{nutritional value}} \times 100$	= ______
3. Convenience score: $\frac{1}{\text{time (min.) for preparation}} \times 100$	= ______
4. Cultural value score: % intent to purchase X 100	= ______
5. Satiety value score: % concept liked overall X 100	= ______
6. Consumer acceptance score[2]: taste panel score X 100	= ______

1 Nutritional value score: % MDR or RDA provided by one serving of product.
2 Consumer acceptance score: Taste panel score (10 highest, 0 lowest) carried out with significant number of people in desired consumer group.

By combining the second set of product dynamics with the first set, he developed the information found in Table 4.7.

TABLE 4.7

MATRIX ARRANGEMENT FOR TWO SETS OF BREAKFAST PRODUCT DYNAMICS FOR TRADITIONAL FOODS

Food Items	Cold			Hot		
	Convenience	Nutrition	Taste	Convenience	Nutrition	Taste
Solid						
RTE Cereal	X	X				
Fruit		X	X			
Doughnuts/ sweet rolls	X		X			
Butter		X	X			
Margarine	X	X	X			
Jam/jelly	X		X			
Toast				X	X	
Eggs					X	X
Bacon					X	X
Pancakes					X	X
Sausage					X	X
Ham					X	X
Potatoes/grits					X	X
Liquid						
Milk	X	X				
Juice	X	X	X			
Coffee				X		X

Source: Diehl (1970).

Combining the two sets of product dynamics provided a tool for showing where new products might be, and what they should be like based on the shortcomings of existing products.

To further identify and evaluate market opportunities he took traditional product data as reported and arranged these into the same product form matrix, as shown by Table 4.5.

Finally, he took a "systems" approach to popular breakfast items, and considered preparation and consumption devices in the same matrix used for the food items (Table 4.8). This blended consideration of product and utensil and provided the final springboard to new product ideas.

TABLE 4.8

BREAKFAST PREPARATION AND CONSUMPTION MATRIX

Product Form	Cold	Hot
Solid	Bowl Plate (small) Plate (large) Knife/fork/spoon	Toaster Small pot Cooking pot Frying pan Waffle iron Broiler Oven Plate (large) Knife/fork/spoon
Liquid	Glass	Coffee pot Tea pot Cup and saucer Spoon

Source: Diehl (1970).

IDEA GENERATION

The point must be made again, that no one individual has a corner on the market for ideas. Crockett (1969) found that so-called creative types are not the only sources for good ideas. He also stressed the importance, at this stage, of using many techniques–approaching the problem areas from different viewpoints to identify solutions or ideas. As an aside, Diehl (1967) found that liquid breakfast items were undeveloped (milk needs more/better taste; coffee has no nutritional meaning).

In summary, some of the more accepted techniques are consumer group sessions, consumer depth interviews, brainstorming sessions, consultants, and published data sources.

It is worth repeating that good new product ideas may arise from various sources on a random basis, many coming from outside a sphere of current interests.

Our own experiences reveal that they often can be triggered by totally unrelated exposures. Crockett also found that, in the past, random ideas contributed to a pool were often the most important idea sources simply because a more disciplined approach did not exist. And even with such approaches, most people believe that random ideas will not diminish in their importance.

Therefore, since the flow of random ideas must be encouraged, it is most important that the people most involved in new product development encourage others to continue to propose ideas. This will provide a continuous input of random ideas into the group pool.

The NIH Syndrome and Its Cure

One of the most common difficulties to be faced in managing new product activities concerns the source of ideas. The NIH syndrome, as it is often called, is the "not-invented-here" and the "not-important-here" complex.

Occasionally there are also problems of another type in having people contribute ideas. Some tend to hold back unless they are certain they will receive recognition as being the source of the idea; or because it is the only idea they think they have and fear being without a good idea when called upon by a superior to help. However, experience reveals the opposite to occur. The revelation of one idea only finds many more appearing, and the flow often overwhelms such people due to the excitement it generates in themselves and others.

NIH is one of the greatest deterrents to progress and is encountered in the most unexpected places and people. A constant vigilance is required to control it. The cure for it and many other similar complexes is contained in an old idea carried out in management discourses: it is surprising to find what can be brought about if one does not care who gets the credit.

EXISTING PRODUCT OPPORTUNITIES

No single company, regardless of its determination, energy, imagination, or resources, is big enough or solvent enough to do all the productive "first" things that will ever occur in its industry and to always beat its competitors to all the innovations emanating from the industry.

But by keeping a close watch on new products being developed by other companies, most companies remain in close contact within an industry. If only new products were developed, and imitation was considered wasteful, then all new products would be end products in the sense that no other company would look at them and try to develop them further. Since most companies have one eye on the market for new products and another on existing old line products, it appears as though both types of products will continue to be developed by companies. The result of this is a successful series of products that have been tested and researched many times over, which vastly increases the excellence of product lines available on the market. Idea generation should also embrace potential imitation product opportunities. Let us explore this in more depth since, in the present day, innovation is more glamorous than imitation, whereas imitative innovations are of equal importance.

VALUE OF ME-TOO PRODUCT INNOVATIONS

Levitt (1966) has suggested that before all our energies and imagination are too one-sided in their direction toward the generation of ideas for true innovations, it is useful to look at the other facts of commercial life. He correctly asks: Is innovation all that promising? Is it all that profoundly liberating? More important, how does a policy of innovation compare in promise to more modest aspirations?

He found that in spite of the extraordinary outpouring of totally and partially new products and new ways of doing things that we are witnessing today, by far the greatest flow of newness is not innovation at all. It is imitation. Further, imitation is not only more abundant than innovation, but actually a much more prevalent road to business growth and profits, he reports. In fact, imitation is endemic and genuine innovation is scarce.

Levitt finds this greater abundance of imitation is perfectly understandable, since each innovation sparks a wave of eager imitators. By the time a so-called "new" product reaches widespread use it has usually been on the market for some time. He believes its visibility is less a consequence of its actual newness than it is of the number of its strident imitators. The newness of which consumers become aware is generally imitative newness, not innovative and timely newness.

Distinctions

Innovation may be viewed, according to Levitt, from at least two vantage-points: (1) newness in the sense that something has never been done before, and (2) newness in that it has not been done before by the industry or by the company now doing it.

The importance of these differences in the character of the required effort and commitment calls for a careful examination by companies of their competitive and growth strategies.

BALANCE NEEDED

The quest for innovation—particularly in new products, in new product attributes, and in customer service—is part and parcel of a company's being marketing oriented. Hence, to have a company style or posture that seeks out opportunities for innovation (big massive ones, or modest innovations to extend the life cycle or broaden the market of a mature product) can make a great deal of sense. Innovation can be one of the most effective possible means of building a company image of progressiveness and leadership, according to Adler (1967).

In an age of explosive science and rapid consumer acceptance of newness, opposition to innovation is a mark either of irretrievable naivety or of helpless blindness. What is needed is a balanced view. Innovation is here to stay; it is necessary, but it does not exhaust the whole of reality. Every company, Levitt suggests, needs to recognize the impossibility of sustaining innovative leadership in its industry and the danger of an unbalanced dedication to being the industry's innovator. He believes that no single company will always beat its competitors to all the innovations emanating from the industry. Further, no single company can afford even to try to be first in everything in its field. The costs are too great; and imagination, energy, and management know-how are too evenly distributed within industries. While true, not everybody clearly acts this way.

REVERSE R&D

Once we become aware that the possibilities of innovation within any one company are limited, we quickly see that each organization is compelled by competition to look to imitation as one of its survival and growth strategies. Imitation, thus, is not just something which even the best managed company will have to be involved in; it is something it will have to practice as a strategy.

This means a company must actively engage in what Levitt (1966) terms reverse R&D. It must try to create its own imitative equivalents of the innovative products created by others. Moreover, the faster the rate at which entirely new products are launched in any field, the more urgent the need for each company in that field to develop a strategy to guide not just the business judgments which must be made, but also the way in which the reverse R&D commitments are made.

The survival and growth of individual companies dictates that they quickly imitate the innovator's new products. Since the speed of competitive imitation so commonly seen tends to quickly cut the margins available to all competitors, the speed with which an imitator enters the market is crucial (Hopkinson 1964).

Surprisingly, Levitt found that strongly new-product-oriented companies, whose products generally required 1 to 3 yr from original idea conception to subsequent market launching, rarely had any kind of policy to guide its responses to the innovations of others. Not a single one of these companies had even given any systematic or sustained thought to the general notion whether it might be useful to have some set of criteria for making commitments to reverse R&D.

In other words, while the companies did a very careful job of planning new product innovations, they had no criteria for the crucial job of new product imitation. Reverse R&D was neither a planned nor careful process.

Levitt's analysis indicates that imitation of a proven product does not automatically reduce the risk; it merely changes its character. While the innovator faces the risk of his product not finding a ready market, the would-be imitator faces the equally hazardous risk of reaching the market when it is already glutted with many competitors. Obviously, the imitator who can substantially shrink his development period below that of other imitators can gain a tremendous advantage. He will encounter fewer competitors and higher and more stable prices during his lead over other imitators. Therefore, not only what the imitator has planned is of consequence, but also what his competing innovator has planned. Idea generation, thus, might also be applied in reverse.

STRATEGY FOR INNOVATIVE IMITATION

For simplicity, let us assume that the genuinely new product of a competitor turns out ultimately to be successful, following the more-or-less classic life-cycle curve discussed earlier. The product is issued at time "0." Most companies quickly become aware of its existence, according to Levitt (1966).

When the innovation is first seen, the pattern in many firms may be similar to that shown in Table 4.9.

TABLE 4.9

POSSIBLE COMPANY REACTIONS TO INNOVATIVE PRODUCT IN FIELD OF VESTED INTEREST

Year	Consumer Response Estimate (%)	Typical Response	Success Probability Estimate (%)
0	0	Doubt it will sell, keep an eye on it	5
1	0	Surprised product still on shelves	10
2	0	Getting a bit more business	20
3	0	Better take closer look	40
4	100	Crash program initiated to imitate	80
5		Reach market, along with several other firms	

Source: Levitt (1966).

Looking back on what happened, Company X, in effect, said at Year 0 that the chances of success for this product were 0%. The judgment made by Company X was zero probability because nothing was done in any positive way to get ready and launch an imitative product. Had the success probability estimates been something above zero, some imitative steps might have been called for, even if only tentative ones. Idea generation machinery could have been organized accordingly, he suggests.

The reason Company X gave a zero probability to the product's chances of success even in Year 3 is that no steps were taken to begin to tackle the job associated with imitation–reverse R&D. Nothing was initiated in the one area that would take the most time and effort if the product were ever to be made by Company X (Levitt 1966).

Probability Estimates

Most people are seldom sure of a genuinely new product's commercial fate. Nobody can ever be completely confident at the outset, or at Year 1, or Year 2, that a competitor's innovation will fail (Adler 1967).

Levitt believes this attitude of doubt should be translated into sound business practice. Suppose that for every genuinely new product issued by a major innovator in its field, Company X required an honest and carefully thought-out coefficient estimate of its probable success–success by some measure of, say, unit sales volume. Hence it might in this case have come up with estimates of the chances of success, at the successive intervals at which it was required to make a judgment that might be like that shown in the right hand column of Table 4.9.

Let us call each of these judgments a Success Probability Estimate (SPE). Suppose now, also, that at Year 0 the policy of Company X was that it must obtain a rough estimate of reverse R&D involved in developing an effective imitation of new product. It could have initiated idea generation procedures earlier. The secret obviously is to start as early as possible while competitors are few and margins are attractive. Thus by Year 3, about 40% of the required reverse R&D would have been initiated, using the Levitt system.

The rate of genuine new product introduction in some industries is obviously too great to justify a company's instigating reverse R&D on each new product that comes along. On the other hand, not every company has ambitions to cover every product opportunity. This, therefore, calls for criteria to select those new competitive products for which reverse R&D should be undertaken and those for which none should be allocated, he believes.

Such a policy might then stipulate that the available resources for this purpose be distributed among the proposals with the highest SPE at Year 0.

CONCLUSION

Idea generating systems exist. They can be initiated at will. Innovation, new product development, extending the life and expanding the markets of existing products by adding new features, styles, and packaging–all these require ideas through which a modern company competes. Innovation is an abundant commodity although probably less abundant than many of us assume. No single company can be as constant an innovator as it is compelled to be an imitator. While there are recognized risks to innovation, there is an equivalent risk in imitation.

Since we live in an age of often very justified faith in the virtues of innovation, it is well to be aware of the possible negative consequences. A key negative effect may be the creation of an environment in which people who frequently suggest imitative ideas are viewed somehow inferior or less worthy; people may then systematically refrain from fostering the imitative ideas upon whose early implementation the continued success of their companies depends.

Supporting a strategy of imitation in some organized fashion in idea generation would have the virtue not only of getting necessary imitative activities into motion early, but of communicating to the entire organization that while innovators are valued, so are the creative imitators.

It makes sense, therefore, to have just as clear and carefully developed a plan for innovatative imitation as for planning innovation itself.

The amazing feature of new idea generation systems is that the process can be made to occur on either innovation or imitation.

In the next chapter, we will examine how ideas are evaluated and the nature of market research.

BIBLIOGRAPHY

Adler, L. 1967. Systems approach to marketing. Harvard Business Rev. *45,* No. 3, 105-118.

Crockett, J. G. 1969. The new product manager in product development. Food Technol. *23,* 26-30.

Diehl, R. K. 1967. Market segmentation vs. positioning in product analysis. Food Prod. Develop. *3,* No. 5, 46-98, 64.

Diehl, R. K. 1970. The method approach. Food Prod. Develop. *5,* No. 6, 26-31, 42.

Eckler, A. R. 1970. Profit from 1970 census data. Harvard Business Rev. *48,* No. 4, 4-8.

Hopkinson, T. M. 1964. New battleground–consumer interest. Harvard Business Rev. *42,* No. 5, 113-119.

Lavidge, R. J. 1968. Equating market opportunities with your product development potential. Food Prod. Develop. *2,* No. 5, 28-31.

Levitt, T. 1966. Innovative imitation. Harvard Business Rev. *44,* No. 5, 63-70; Sept.-Oct., 119-126.

Lorsch, J. W., and Lawrence, P. R. 1965. Organizing for product innovation. Harvard Business Rev. *43,* No. 1, 109-122.

Mainer, R., and Slater, C. C. 1964. Markets in motion. Harvard Business Rev. *42,* No. 2, 44-49.

Reynolds, W. H. 1965. More sense about market segmentation. Harvard Business Rev. *43,* No. 5, 107-114.

Steinberg, C. S. 1958. The Mass Communicators. Harper & Row, New York.

Yankelovich, D. 1964. New criteria for market segmentation. Harvard Business Rev. *42,* No. 2, 58-71.

CHAPTER 5

Market Research

From a marketing point of view, a new product emerges in four phases: a search for the idea, translation of it into a business proposition, development of a marketing plan, and validation of the plan. In the last chapter we explored idea generation. In this chapter we will examine market research, the system being used by nearly all companies and agencies, in one form or another, to carry an idea forward into the market place. The entire sequence of events is shown in Fig. 5.1. The elements of market research are summarized in Table 5.1.

OBJECTIVE OF MARKET RESEARCH

Jenkins (1970) noted that during the decade of the fifties, market research as a tool of management literally exploded. Suddenly everyone discovered the probability sample from political pollsters to the broadcasting industry, from automobile companies to pet food manufacturers. Everything from relatively small sample motivational studies to enormous omnibus studies were the order of the day.

Jenkins believed this happened because it was a period of great economic growth and expansion. Large capital investments were being made with equally large marketing appropriations to support their productivity. Hence, management turned to market research as a panacea to minimize its risks by predicting future behavior. As it was then, it still is today. He stated that the objective of market research was and is simply to minimize risk!

NATURE OF MARKET RESEARCH

He goes on to suggest that marketing people have learned something from these experiences, so that the nature of market research and the manner in which it is practiced today in sophisticated companies is somewhat different. One element of its nature, he believes, is very much akin to baseball: both have generated more data, more statistics whether measured by volume, weight, or the linear foot than any other activities which readily come to mind.

While it is interesting information, is it useful, germane, or actionable? Companies have come to learn that market research is a delicate tool to be approached with caution, and probably know more about what it cannot do than what it ultimately will be able to do in the future, especially through the application of the computer. The unintentional misuse and misinterpretation of this kind of work in the hands of unseasoned marketing people can lead to disastrous results. In addition, over the years another element has crept into the nature of the use of market research, according to Jenkins (1970). It often becomes a crutch in the hard task of judgmental decision-making—an arbiter between two conflicting points of view, but subject to the individual interpretation of each protagonist to support his own positions.

In much the same way, market research cannot substitute for creativity as it is sometimes asked to do. In the field of new product development, it cannot generate the creative concept. It may not be able to tell what new products people want because they themselves cannot articulate their needs or desires for products or services that have never existed and, therefore, they have never experienced (Jenkins 1970).

Experts in the field believe that market research can provide innumerable insights into the market to assist in the ultimate marketing-management decision by enlightening them on the results of their own activities and, just as importantly, on those of competitors.

It can also reflect consumer attitudes toward product, price, packaging, promotion, and the advertising message; identify areas of opportunity and weakness; measure trends, identify broad markets in terms of geography and all of the basic vital demographics; trace usage in terms of frequency and again in terms of demographics; and evaluate concepts of new, untried, and unborn new products.

IDEA FILTRATION NEEDED

With little effort, most people could find and quote a source which stated within the past year that the rate of failure among new products amounted to between 80 and 90% of new product attempts. The exact figures would vary according to the criteria of success and failure employed and the source cited. As Orman (1967) pointed out, two facts would remain after the figures were quoted, the sources discussed, and the criteria considered. One, there is a belief that new product failures are higher than they should be; and two, that many introductions are not to be found on the shelf when the first anniversary of their introduction date rolls around. Further, some of the products which are still on the shelf on their anniversary date have yet to show a profit to their manufacturer. Most people would also feel that the high rate of new product failures suggests that a great weakness in business today is new product idea evaluation. This notion can be documented.

HIGH FAILURE RATE AND COST

Angelus (1970) reported that in 1968 about 10,000 supermarket items were introduced. Approximately 80% failed. Of the 8000 failures, 7000 were in test markets or regional markets, 1000 were major national failures. During the 1970's, over 120,000 new products will be introduced into the supermarket field. One hundred thousand will fail. The media waste alone on these products will amount to $5 billion on test market products and $7 billion on national products.

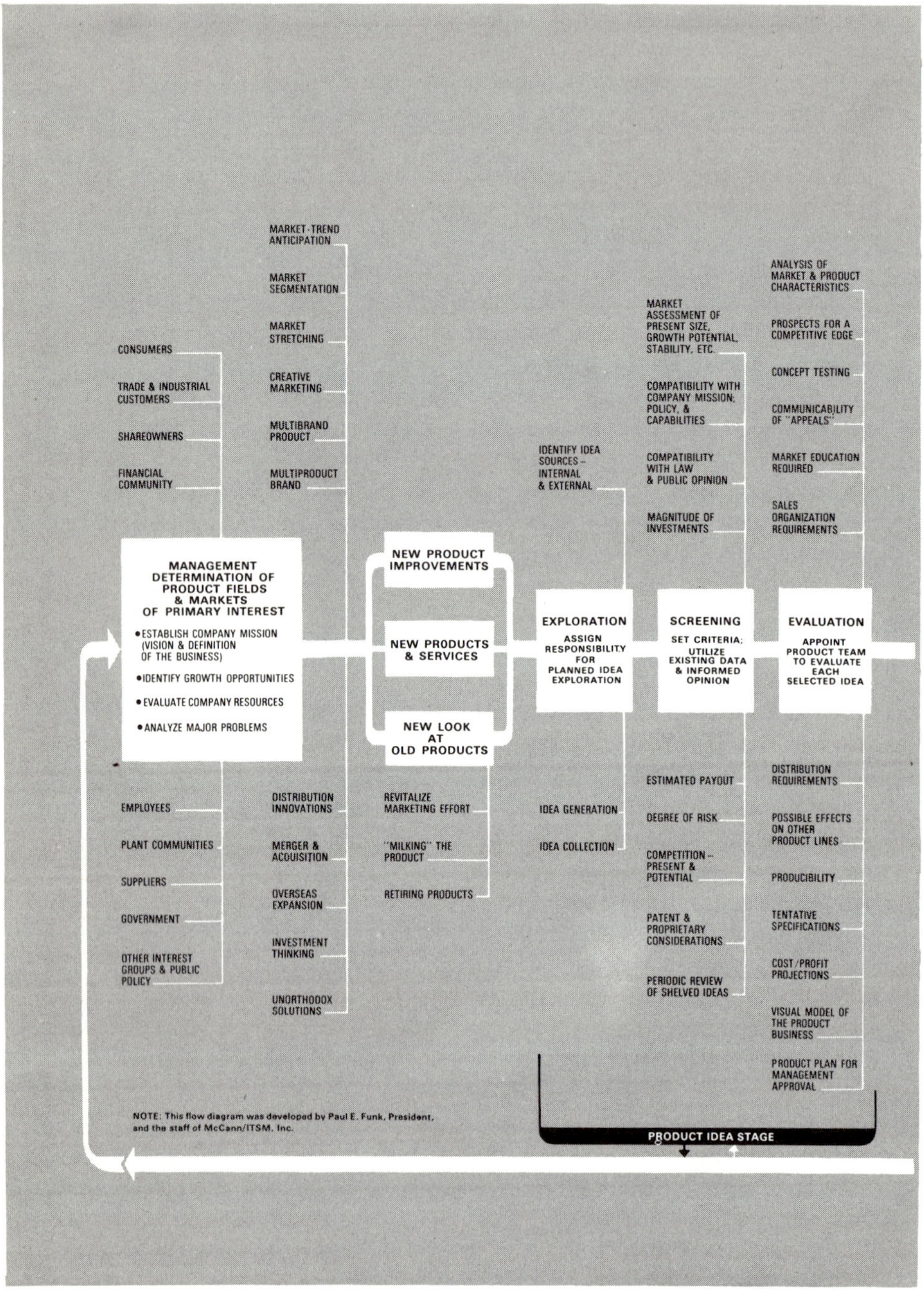

FIG. 5.1. WORK FLOW AND SYSTEMS CHART FOR MANAGEMENT OF NEW PRODUCTS

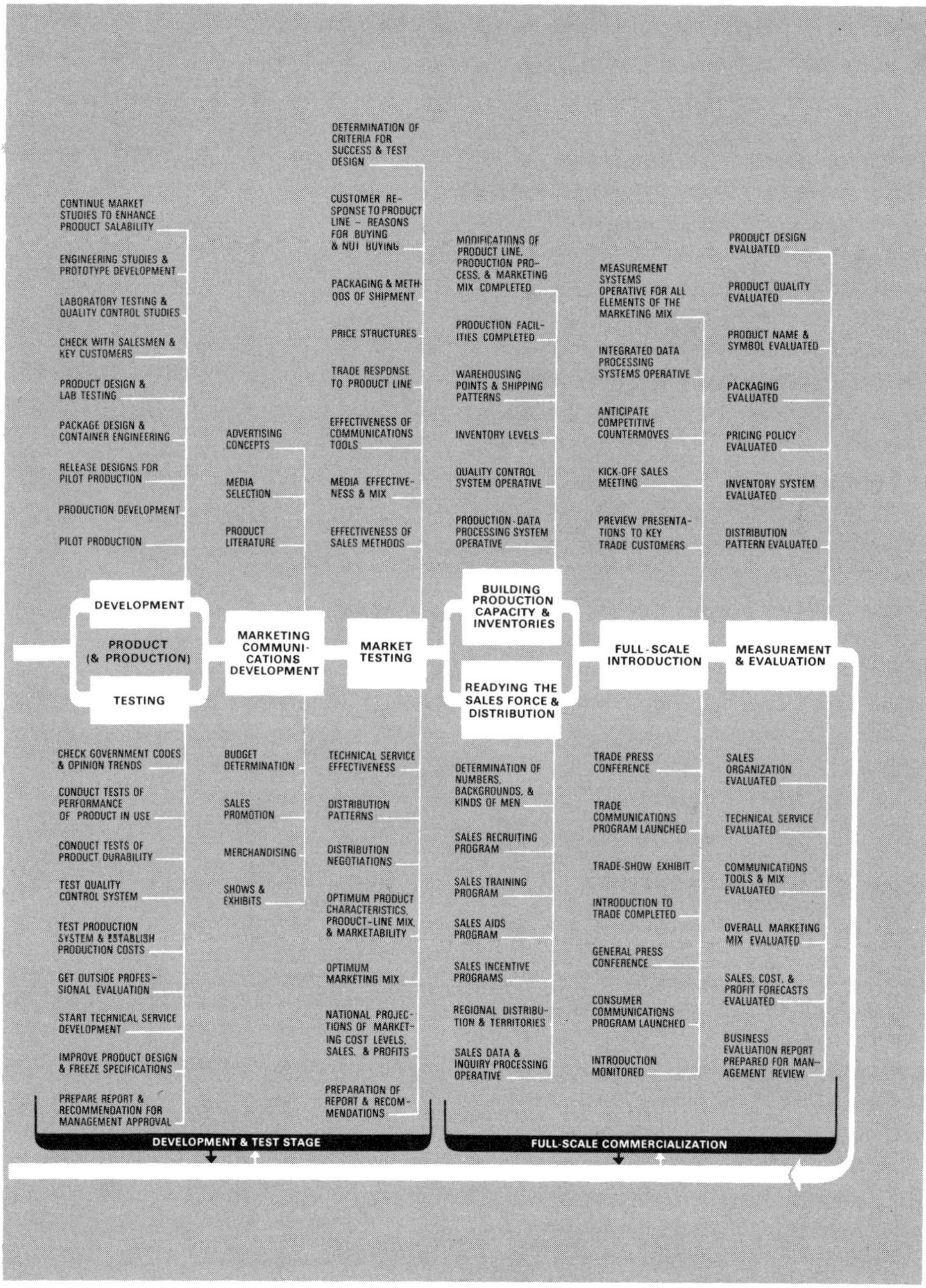

Courtesy of Adler (1967) and the Harvard Business Review

TABLE 5.1

ELEMENTS OF MARKETING RESEARCH

- Idea generation
  - R&D experimentation
  - Outside consultants, inventors, manufacturers, wishing to sell materials, machines, etc.
  - Internal "brain-storming"
    - Synectics sessions
    - Ideas submitted by employees
  - Studies of secondary source data
    - Economic trends
    - Demographic trends
    - Grocery sales trends by category and subcategory
    - Studies of changing life styles
    - Availability of new technologies
    - Competitive activity
    - New product introductions by other companies
  - Computer-based product development procedures
  - Analysis of consumer needs in specific product areas

- Idea assessment
  - Availability of production facilities, technology, distribution system, etc.
  - Preliminary P&L, margin potential, etc.
  - Group sessions utilizing concept descriptions, prototype product, etc.
  - Individual depth interviews
  - Quantitive concept of product and concept testing

- Product assessment
  - Taste tests
    - Double triangle
    - Paired comparison
    - In mobile trailers
    - In shopping center store
    - In supermarket
  - In-home use tests
    - Monadic blind tests
    - Paired comparison blind tests
    - Sequential monadic blind tests
    - Product alone (blind)
    - Product with concept
    - Identified product
    - Alternate formulations
    - Versus competition
    - Alternate packaging, names, positionings
    - Personal placement and reinterview
    - Mail placement
    - Telephone, mail, telephone single placement
    - Multiple placements over time to set extended use effects
    - Evaluation by housewife only
    - Individual family member evaluations

- Packaging assessment
  - Test alternates, test new versus current, etc.
  - Shelf visibility
    - Tachistoscope
    - Slides
    - Mock-up shelves
    - In-store shelves
  - Specific reactions and attitudes
  - Tests of alternate constructions, shapes, etc.
    - In-home use tests

Group sessions
Trailer tests

Advertising assessment
Qualitative developmental
Utilize storyboards, print ads, animatics, rough commercials, etc.
Group sessions
Small scale comprehension and communication tests
Quantitative evaluation
Test exposed versus control
On-air television test using rough commercial or finished commercial
Shopping center location
Trailer
Theater

Assessment of total
marketing effort (product, package, concept, advertising, price, etc.)
Prior to full-scale test market
Purchase simulation
Extended in-home use and purchase
Controlled store tests
Mini test market
Full-scale test market
Distribution checks
Store audits
Awareness, attitude, and usage studies
Consumer diary panels
Tests alternate marketing mixes: promotions, media plans, pricing, shelf location, etc.

Source: Baen (1970).

The cost of failure will be another major new product problem in the 1970's. Angelus found that a modest test market failure can cost approximately $100,000; a nationally introduced product that fails in the market may cost 5 million dollars or more. The above being true, industry is going to waste many R&D expenditures in new product work over the next 10 yr just as it did in the last 10 yr.

At the same time development time will increase. A minimum of six months will be needed to introduce a minor line extension product without technical obstacles to overcome. As many as 3-5 yr may be necessary in developing a product having a major technical innovation. There is a real danger that new products will be obsolete before they are introduced.

Each of these problems is critical. The most critical one, however, is the high failure rate of new products. If one can achieve a high success ratio with new products, the other problems become minor.

In analyzing the products that failed to meet desired sales goals, over 150 executives from manufacturing, marketing, advertising, and research areas were interviewed by Angelus. The major results are summarized in Table 5.2.

HUMAN VALUES A FILTER CRITERION

Common to expert thinking on the subject of new product ideas is the notion

TABLE 5.2

REASONS FOR FAILURE

Reasons	%[1]
Vague consumer difference	36
Poor product positioning	32
No point of difference	20
Bad timing	16
Poor product performance	12
Wrong market for company	8

Source: Angelus (1970).
[1] Total may be more than 100% because of more than one reason.

that we need some filtering device which will reduce the number of ideas to be worked on in some selectively positive fashion. The fabric of this imaginary filter is woven of human values.

Orman (1967) reported that at least five commercial research organizations practice explicitly some translation of the recognition of human values as being interposed between the consumer and the objects available to her.

Defining the Classes of Values

We all recognize that human values are intangible, which literally cannot be directly felt, handled, tasted, or smelled because they are not objects. Yet we all deftly relate our perceptions of these values to respond accordingly. As consumers, we each are interested in what a product represents to us and what it promises to do for us. Since new products now are usually presented to us through advertising, which is an abstract argument to create interest in a product, our first purchase will be based upon perceived values.

Orman (1967) reported that behavioral scientists say human behavior is goal-directed, and that consumer behavior is a specialized form of human behavior, and, as such, is thereby goal-directed. He suggested that in this theory of goal-directed behavior the anchor points are human values; goals are selected because the goal seeker possesses and is guided by her values. In the supermarket, the consumer is acted upon by many forces. Her system of values selects out the alternatives presented which she will listen to, which she will act on, and which she will seek to satisfy.

Orman and others believed that when the consumer perceives a product, it is seen as something more than an object of certain physical dimensions. There is a translation into value components and these components are considered in a cluster. For analytical purposes, Orman suggests that the cluster is composed of the following value elements:

Structural.—How good is the product quality? How much will it cost? What alternatives are available?

Functional.—Will the product perform? Does it provide a sound price-value? Will it be convenient to use? Will it be eaten by family members to whom it is served? Will it be satisfactory as a leftover?

Orientation.—Is it characterized to be traditional, modern, exotic, mundane, dominant, etc.? Each relates to an internalized value held by the perceiving individual which ranks high or low in her selection process.

Symbolic.—What do the symbols associated with the product mean in the world, and how are they to be used to enhance one's position within this world? Brands and styles are examples of such symbols.

Emotional.—What aesthetic qualities does the product have? (How it looks, how it will taste, smell, and feel, etc.)

In summary, Orman states that there is a cluster of component values with which a human will react when responding to an object. To make that object meaningful, we will filter our observations concerning that object through the system of values we possess. Any resultant evaluation and intention to act in regard to that object will be in terms of how that object fares in relation to our value systems. To improve a product's chance of success, a product's concept components should be sifted through the relevant portions of consumer value systems for approval or disapproval.

CHANNELED CREATIVITY

Jenkins (1970) summarized the modern corporate view. He reported that in today's competitive environment new products offer management one of the greatest opportunities for the maintenance of corporate profits and the provision of avenues for growth. In the hue and cry and pressure to obtain new products, we can also be misled. The ultimate criterion should be "Does it fit?" Does it fit the corporation's capabilities in terms of our resources of money, manpower, and expertise to enable the corporation to gain the competitive edge? Does it fit consumer needs?

To answer one factor in the question "does it fit," we need methods of screening new product potentials. Experience reveals that if these are carefully examined and defined, according to consumer values, there is a major probability of success. Review and analyses reveal that successful approaches include the following: (1) Channel the best ideas through a filter to determine whether they will gain initial consumer interest. (2) Research product concepts to select the most promising for development. (3) Follow this with prototype product development. (4) Retest the concept with prototype products, and be guided accordingly.

CONCEPT VERSUS PRODUCT TESTS

The next step is to test the concept generated. While there are varying opinions, the consensus is that testing concepts rather than committing physical execution of development at this stage of exploration has advantages. Orman (1967) lists the following reasons:

More Flexible

Concept testing is more flexible than product testing. Verbal descriptions permit wide expression for a concept while a physical product affords only minor variations. Within the structural, functional, orientation, symbolic, and emotional value system described earlier there is a multiplicity of potential combinations. Flexibility also extends to the immediate redirection and revision of concepts as data become available.

More Economical

Many ideas can be evaluated in a single screening; a dozen or more can be managed efficiently.

More Immediate Results

Idea testing can be brought about faster than product testing since prototypes are not needed. Concepts can be thoughtfully prepared in hours. Product prototypes may take weeks to develop.

STRUCTURING CONCEPTS

Experience in idea screening reveals suggestions for improving results.

Written concepts for foods are best framed in sensory, nutritional, socio-psychological, economic, health, and convenience terms according to Orman (1967).

Once concepts have been written, they may be refined by exposing them to several small groups. This step is optional and depends largely upon the clarity with which consumer needs have been expressed in the concept generation which preceded the writing of the product concepts. The most effective interviewing is conducted within consumer homes. Interviewees are usually presented with a card, on which is written a product-concept to be screened. A number of questions are asked about what she perceives. Respondents can usually handle 3-5 concepts, 1 at a time, without confusion. A hundred respondents provide an adequate base. That number generally contributes all the insights needed to reach conclusions on the concept (Orman 1967).

How do competent groups approach market research? Analysis reveals the following sequence of market research practices at work.

The Idea Search

This step involves identifying opportunity areas, generating ideas, and evaluating ideas. The overall sequence is shown schematically in Table 5.3. In previous chapters, market opportunity areas and idea generation were discussed in some length. We now turn to idea evaluation.

TABLE 5.3

CRITERIA FOR EVALUATION OF IDEAS

Marketing variables
Annual dollar volume potential
Uniqueness of concept
Incremental product value
Estimated life cycle
Competitive communication
Consumer purchase situation
Feasibility consideration
R&D capabilities and facilities
Production capabilities and facilities
Uniqueness of process
Identity and regulatory

Source: Crockett (1969).

Specifically, each idea must be examined in the context of its potential marketability, its feasibility from the company viewpoint, and consumers' reaction. In an initial consumer screening, Crockett (1969) suggested what an idea and its description should be like:

Idea: Instant Breakfast

Description: A powdered mix which can be mixed with a glass of milk to provide a nutritious balanced breakfast in beverage form.

The consumers in a panel would be asked to say "yes" or "no" in terms of interests, and then asked to give direction if the idea has some merit. It is important that only the obviously uninteresting ideas are eliminated. Following this initial screening, the remaining ideas are then evaluated against preestablished marketing criteria and company considerations, such as shown in Table 5.3.

With this screening process completed, ideas which have the best potential commercial value and are most relevant to the company are now ready for further testing to obtain consumer reactions in greater depth. This is shown as Stage No. 5 in the 5-Stage Funnel presented in Table 4.1 (Chap. 4).

Concept Testing.—The testing technique, according to Jenkins (1970) is designed to develop consumer reactions to a particular new product idea by showing them a short statement describing the new product idea in detail. In Crockett's example, the idea, Instant Breakfast, can be expanded into a full-fledged concept statement such as the following:

"Now you can have a complete breakfast in a new convenient form. One glass of Instant Breakfast provides all the vitamins, proteins, and other healthful nutrients contained in two eggs, bacon, toast and butter, and 4 oz orange juice.

"Simply mix the contents from the package with 8 oz of milk, stir, and it's ready to drink.

"Comes in chocolate, vanilla, strawberry, and coffee flavors. Costs 15¢ per serving (Crockett 1969)."

Analysis of the results of such testing reveals requirements for further refinement of the concept, or the decision to initiate laboratory development.

Idea Translation

The next step involves converting concepts into consumer-acceptable products and translating them into acceptable propositions as well.

This phase involves the steps by which concepts are converted from words and ideas into products that are acceptable to the consumer and fulfill her needs; it also includes a business proposition acceptable to company management.

At the conclusion of the search phase, the market researcher gathers and evaluates the data. From this analysis he identifies those concepts to develop into new products and put on the market.

This information, Crockett explains, becomes the main selling tool to company management, and takes the following form: (1) Description of the concept. (2) Market appraisal. (3) The opportunity. (4) The preliminary business proposition. (5) Formal project request.

While the specific forms vary from company to company, the basic document, he found, also provides the initial request for R&D development. The laboratory specifically requires the following data: (1) The concept in detail. (2) The consumer and market appraisal. (3) Volume goals. (4) Timing schedule desired. (5) Specific product specifications. (6) Packaging guideline.

At this point, a work or venture team will usually be organized in the company to supply functional management in the various development phases. The team leader must remain in control throughout the project, and can probably coordinate the many variables more effectively if he utilizes some variations of the PERT system approach, to be discussed later. If the above is successful, the next step is to convince management on a business basis of the soundness of the venture because major resources will be required to bring the product to test market, regional market, and then go nationally.

The major elements of a business proposition are the following:

(1) An appraisal (current and future outlook) of the market situation we are proposing to enter, the voids we see, and the concept we propose for filling the void.

(2) The consumer's viewpoint regarding the void (or problem area) and her reactions to our proposed solution (the product and concept).

(3) The marketing plan required to achieve targeted levels of awareness, trial and acceptance of the new product, and the level of investment required to maintain repeat usage.

(4) The translation of the marketing plan into a business proposition with pro forma P&L's over the life cycle of the new product, sequential capital invest-

ment requirements, a cash flow, the break-even point, and returns on assets employed and/or investments (Crockett 1969).

Plan Formation

If all goes well, the next step involves developing the market program required to achieve goals established in the business proposition, with a plan for testing it on a limited basis. Experts in the field agree that the planning stage involves: (1) The development of the marketing program required to achieve the goals established in the business proposition and approved by management. (2) A plan for testing the market program on a limited basis to minimize the risk. This area will be covered in more detail in Chap. 15.

Plan Validation

The final step is to validate the above plans. This involves actual implementation of the test marketing plan, modifying it as required, and, when goals are achieved, turning the successful new venture over to an operating division. The team members then usually start over, in searching out another new product, or developing a family of line extensions for the newly born product, or profit center, depending on the company situation.

In the use of a venture team for the activity, the team members can become an operating profit organization of their own in a corporation, and draw accordingly. This will be revisited later.

A summary, outlining the basic elements of market research, is presented in Table 5.1. The relationship of laboratory development to marketing research development of a new product is shown in Table 5.4. While there is a meshing of activities, there are also distinctive differences which are apparent. How these differences are managed and brought together will be the subject of subsequent chapters.

TABLE 5.4

RELATIONSHIP OF PRODUCT DEVELOPMENT AND MARKETING RESEARCH

Product Development	Marketing Research
I. Idea Generation and Refinement	I. Idea Generation II. Idea Assessment
II. Technical Development	III. Product Assessment IV. Packaging Assessment
III. Pilot Plant Scale-Up	V. Advertising Assessment VI. Assessment of Total Marketing Effort

Source: Baen (1970).

BIBLIOGRAPHY

Adler, L. 1967. Systems approach to marketing. Harvard Business Rev. *45,* No. 3, 105-118.

Angelus, T. L. 1970. Improving the success ratio in new product development. Food Technol. *24,* No. 4, 29-36.

Baen, J. 1970. Relation of product development to market research. Meeting in Miniature. N.Y. Inst. Food Technologists, Jan. 20.

Crockett, J. G. 1969. The new products manager in product development. Food Technol. *23,* No. 6, 26-30.

Ferrell, R. W. 1964. Customer oriented planning. Am. Management Assoc., New York.

Jenkins, F. A. 1970. The culmination of product development and market research in consumer communications. Food Technol. *24,* No. 4, 60-67.

Katona, G. 1964. The Mass Consumption Society. McGraw Hill Book Co., New York.

Levitt, T. 1965. Exploit the product life cycle. Harvard Business Rev. *43,* No. 6, 81-94.

O'Meara, W. H. 1961. Selecting profitable products. Harvard Business Rev. *39,* No. 1, 95-102.

Orman, A. D. 1967. A systematic approach to screening new product ideas. Assoc. Natl. Advertisers Workshop on New Products and Services. Plaza Hotel, New York, May 10.

CHAPTER 6

R&D Communications in a Mass Marketing Company

At its best, a research laboratory is a window into the future. The depth and height and breadth of its vision is, in most cases, related to the nature of the organization within which it exists, and how it is brought to bear on pressing company issues.

In the old days, some company managements saw their R&D organizations as a collection of "long hairs" maintained for status purposes. As such, they might have come up with something of value, but probably not. Another old company attitude was to relegate R&D to trouble shooting and control work. While these are required functions in any company engaged in manufacture, they obscure the basic reason for R&D in any modern, well-managed enterprise. Crisply stated, this is to increase current profits and to insure future profits.

Top. management now recognizes its responsibility and takes time to understand R&D as it does the other phases of business. Presented in this chapter are methods for the required communications and controls, as well as allocations, of responsibilities for various aspects of the production and marketing effort in new product development, maintenance of older products, as well as a discussion of related aspects of Research and Development.

R&D DIVISION IN THE COMPANY ORGANIZATION

Inasmuch as R&D is essential for the viability of a company, the division should report to the chief executive officer. This is important for direct communication and, further, places the division on an equal basis with other segments of the business. The participation of the business head in policy-making decisions on various problems involving R&D is facilitated by this organizational approach.

R&D. IN A MASS MARKETING COMPANY

In the following, it is assumed that production control operations are placed in the production division. It is further assumed that preparation and issuing of manufacturing standards and control methods to be used will be the responsibility of R&D. Below is a listing of some of the many responsibilities of R&D in a mass marketing enterprise.

(1) Development of new products.
(2) Maintenance of existing products in a good competitive position.
(3) Continuous review of changes in competitive products as requested by the Marketing Division.

(4) Regular examinations of company products as bought in retail channels.
(5) Development of processes for new products.
(6) Improvement of processes for existing products.
(7) Preparation and maintenance in an up-to-date condition of manufacturing standards for all company products.
(8) To keep all company products in a competitive position cost-wise.
(9) To review product complaints as a guide for needed changes in product or process.
(10) To assist production in the start-up of new or improved processes as requested.
(11) To develop rapid and cheap analytical methods for use by the control and research laboratories.
(12) To compile and issue up-to-date methods for control of the purchase of ingredients and control of production.
(13) To institute a procedure for establishing the accuracy of control analyses by the regular distribution of check samples.
(14) To determine adequacy of all packaging materials for protection of product.
(15) To assist other divisions on request in design changes of packaging materials.
(16) To approve the technical accuracy of proposed advertising claims.
(17) To establish safety of all company products under U.S. FDA and U.S. Dept. of Agr. requirements, or appropriate agencies.
(18) To insure that competitive and company samples used in consumer surveys are representative of normal quality in trade channels.
(19) To maintain samples of all consumer survey material and obtain necessary analytical and other data to aid in interpreting survey results.
(20) To accumulate background information on products of companies under consideration for acquisition.
(21) To assist in selecting consultants to handle special problems for which manpower and facilities are not available in the Research Division.
(22) To establish and maintain regular communications with the Marketing, Market Research, and Production Divisions of the company. Communications with other segments of the business should be on an "as required" basis.
(23) To maintain contacts with scientists, societies, suppliers, research groups, and others helpful to the company.

METHODOLOGY FOR EFFECTIVE FURTHERANCE OF THE MOST IMPORTANT R&D FUNCTIONS

The one most important factor in resultful R&D operation for a company is a regular meeting, probably monthly, comprising the chief executive officer, R&D head, marketing head, and production head. As will be noted later, this "Research Committee" may be expanded somewhat if desired, but limitation to individuals who have responsibility for effecting some portion of decisions reached is advisable. This meeting also serves as a communication medium on the research program as it impinges on the several segments of the organization responsible for the commercialization of the R&D results.

NEW PRODUCT DEVELOPMENT

This is one of the most important activities of R&D and should proceed in an orderly fashion when properly organized. This will be discussed in detail in subsequent chapters. Nevertheless, it should be understood at the outset that as the development of product is going on, process development, packaging studies, probable shelf-life, and other related factors will have been instituted in R&D at appropriate times. This is an area where regular meetings of the Research Committee are of particular value for communication, advice, and decision.

MAINTENANCE OF COMPANY PRODUCTS IN A GOOD COMPETITIVE POSITION

The profitable life span of a company product on the market is always a problem since competitive developments continually force reappraisal of its consumer acceptability. This appraisal generally is made by a blind consumer survey against principal competitors' products. It is, of course, preferable that the company be a leader in innovations rather than a follower. Research, therefore, has the twofold responsibility of modifying products to equal or better competition and to strive to lead in desirable product changes and, thereby, force competition to follow. It should be noted that relatively small changes in product or packaging may result in consumer advantages sufficient to support an advertising program leading to increased sales, longer product life, and increased profitability.

The revitalizing of products on the market should not be taken lightly by the Research Division because, on the average, the life expectancy of a new product is today much shorter than it was even 10 yr ago.

This means that the permissible time for recovering R&D expense and introduction costs is shorter. Accordingly, the advantages of maintaining a superior competitive position are clearly evident.

REVIEW OF COMPETITIVE POSITION

A continuous flow of authoritative information about the composition and properties of new products introduced by competitors and modifications in their brands currently being marketed is essential for the marketing divisions of the company.

The marketing group generally will become aware of these changes; frequently before the new product appears on the shelves. They can procure samples of new products and old ones which have been "improved" to maintain or increase sales and send these with all helpful information to R&D.

Generally, the determination of composition and physical examination of older products will suffice for a marketing division estimate about how the change will effect their sales of similar brands. Further action in the direction of changes in company products will be dependent on this evaluation.

Marketing Division analysis of the R&D data accumulated on new products should lead to a decision as to company interest in marketing a similar product. If the answer is affirmative, R&D will need to develop it with celerity.

REGULAR EXAMINATION OF COMPANY PRODUCTS IN RETAIL CHANNELS

Control of production of company brands is the responsibility of the control laboratories which commonly report to the production head.

Human nature being what it is, selection of samples for control observations, or other attempts to make the products they manufacture appear to be of better quality than they actually are, may frequently be made by department foremen and operators. This practice can lead to a false opinion about quality and uniformity of production and may seriously affect the consumer's reaction to the brand with resulting decrease in sales.

Regular purchases by an outside agency, such as Nielsen, of important company brands manufactured in different plants, from chain and independent stores should be inspected by R&D. Results of such regular inspections yield the following items of useful information. (1) The general quality of company brands on the shelves ready for purchase and whether it is about as expected from control laboratory reports. (2) The efficacy of the packaging in protecting product during warehousing, storage, and store handling. (3) Whether there are significant differences in average properties of product among plants. In other words, how uniform are the properties and quality of each product in all segments of the market? (4) How frequently does substandard product escape detection by the control laboratories?

DEVELOPMENT OF PROCESSES FOR NEW PRODUCTS

When a new product reaches the development stage where it appears to be feasible, R&D must begin to plan an economical process for its manufacture. This is accomplished in cooperation with other company elements, as will be discussed in Chap. 7.

In many companies, the production division has an engineering group for design and maintenance of plant. In such cases good communications are essential between Research Engineering and Plant Engineering. Each group can profit from the specialized knowledge of the other.

As an example, Research Engineering will generally accumulate all the data on physical properties required for process equipment selection design. Production Engineering can then lay out the process with due regard to available space, available equipment, and other production considerations.

Research Engineering will generally have worked with small scale pilot equipment to obtain the necessary physical information. Presumably, also, the pilot plant will have used either specially built or available equipment in producing pilot lots of the proposed product. It is well to keep in mind that small pilot operations frequently give differing results from plant operations where there may be a 100- to 1000-fold increase in size.

Once a production process has been designed and installed, the start-up generally requires that both research and production personnel be involved until satisfactory operation is achieved. Incidentally, both divisions benefit from such cooperation.

IMPROVEMENT OF PROCESSES FOR EXISTING PRODUCTS

R&D should be always alert to the possibility of improvement of existing processes either to lead to a better product at no increase in cost or to lead to cost savings without impairment of quality. Here again, good communications between the Production Division and R&D are essential for best results.

PREPARATION AND MAINTENANCE OF MANUFACTURING STANDARDS

Maintenance of up-to-date standards for the manufacture of all company products is an important company function best handled by R&D. These standards should contain all essential data, such as ingredient purchase specifications, product formulae in terms of chemical composition, batch weights, temperature and time conditions for each step in the process, control standards the product must meet, and any other factors vital for the quality of the product.

In general, the standards should give the operators as much flexibility in operation as possible without affecting the quality of the product. The reasons for this lie in the unavoidable variations in processing due to unforeseen factors, such as breakdowns or other processing interruptions. For example, a minimum time and temperature of processing at a given stage may always be essential, but a somewhat longer time may not be hurtful. In this case a minimum and maximum time might be specified.

All standards changes should be promptly handled and issued to the various persons authorized to receive them. Here in the interests of security, one individual should receive only that portion of the standards needed for his own part of the operation. For example, the purchasing head has no need for process and product specifications, but does need specifications for ingredients and packaging materials to be purchased.

Production, research, or purchasing may request changes in manufacturing standards. This may be done on a special form setting forth the need for and details of the proposed change.

Both the production and R&D heads sign each original sheet of any manufacturing standard and likewise any succeeding changes. This insures agreement before issue. All copies will, of course, carry this data.

It is important that standards and standards changes issue promptly. It is also important that all copies of superseded standards be destroyed. One certain way is to recall all superseded sheets to the group in R&D responsible for issuance and have them destroy all but the original copy. This is retained in a historical file which contains the entire story of all changes of any type for the particular product from the time of initiation of the standards system.

The preceding gives a very brief sketch of a manufacturing standards system which, with suitable changes for a particular company, can be an invaluable tool for insuring uniformity of production. Provided that the procedure outlined is rigorously followed, the file of past actions on products is also invaluable for checking back on probable causes of product troubles which may be encountered in trade channels. The code dates on the products in question enable one to find out easily the exact operation which was in use at the time of manufacture.

KEEPING COMPANY PRODUCTS IN A COMPETITIVE POSITION COST-WISE

For commercial success in a mass marketing operation, it is of the utmost importance that company products be kept competitive from the cost standpoint. The one exception is the relatively rare case where a product enjoys a premium price because of characteristics which set it apart from other products of the same type. Such cases may arise from a favorable patent position, superior sources of raw materials, or a captive source of ingredients not available to other manufacturers. Products in such a favorable position are of infrequent occurrence and often the favorable factors are of limited duration. Accordingly, R&D must be alert at all times to insure that insofar as possible, a competitive advantage or, at a minimum, equality of cost is maintained for all products of significance in the company profit picture.

PRODUCT COMPLAINTS REVIEW

R&D should review all significant product complaints as a source of information about product or processing troubles which need correction. The number of complaints about a specific shortcoming will be an index to guide the perceptive research man to specific problems with existing products which need research attention.

ASSISTANCE TO PRODUCTION IN THE START-UP OF NEW OR IMPROVED PROCESSES

Production should have the assistance of R&D as needed when new or improved processes are placed in operation. This is useful because each division has experience and background information which can contribute to a speedy solution of problems incidental to every untried process start-up.

DEVELOPMENT OF RAPID, CHEAP, AND RELIABLE ANALYTICAL METHODS

The compilation of and distribution of analytical methods for use in the development and control of company products, and for the analysis of competitive samples, should be the responsibility of R&D. For control use the methods should be rapid, accurate, and cheap.

Instrumentation for control analysis may be desirable, particularly if the cost savings in manpower are greater than the amortization cost of equipment. Savings of important magnitude coupled with quicker availability of data can frequently be realized by such techniques.

Often, instrumental, chromatographic, or other procedures are needed for research projects and for the analyses of competitors' products. These, as well as control methods, should be developed, recorded in a standard form, and kept up-to-date in an analytical methods book.

ANALYTICAL METHODS FOR THE CONTROL OF PURCHASES AND PRODUCTION

R&D should issue and keep up-to-date methods used for the approval of all purchased ingredients and the control of processes and product. These are for the use of the control staff and should be exactly followed at all times. Since such methods may be used by technicians, they should be carefully and explicitly written to insure understanding by the user. This can be best done by using a carefully conceived format. The methods can best be inserted in loose-leaf form in a Methods Book with a numbering system for easy expansion and indexing.

PROCEDURE FOR ESTABLISHING ACCURACY OF CONTROL ANALYSES

A company, particularly if it has production operations in several different factories, should maintain a fairly frequent check on the accuracy of control procedures. R&D can readily keep check on this by distribution of samples of company products for analysis by the control staff assigned to the particular determination or determinations involved. Naturally, the control methods as employed by the analyst should be followed. From the compilations of results by each of the control analysts, as well as from the research analyses of the samples, a regular check on the adequacy of the control methods and control staff can be readily maintained.

ADEQUACY OF PACKAGING MATERIALS

Packaging development for company products necessarily is a divided responsibility. Marketing is interested in eye appeal on the store shelf and ease of use of product by the purchaser. Production has problems in handling the packaging on production lines without undue damage to the container or wrappers or excessive down-time due to packaging equipment troubles. R&D has the responsibility for determining the adequacy of packaging materials for protection of the product during shipment, warehousing, on the retailer's shelf, and through the hands of the consumer until it is consumed.

Protection of the product implies freedom from physical damage through commercial channels of distribution. It also implies that the packaging material does not itself have any effect on the quality of the product. For example, the materials used in packaging should not adversely influence the color, flavor, or odor during the life expectancy of the product.

This means that any new or changed packaging material should be adequately tested on each and every item for which it will be used.

The testing program must be specifically designed to cover temperature cycling, accelerated aging, shipping tests, or any other type of testing which experience may dictate as useful for uncovering problem areas. Failure to carry out such testing can result in serious product problems in trade channels.

ASSISTANCE TO OTHER DIVISIONS ON DESIGN CHANGES FOR PACKAGING MATERIALS

R&D facilities and knowledge about properties of product and packaging should be available on request to aid other divisions in the development of desired designs and processing utility of packaging. They should also assist in evaluating packaging ideas brought to company attention by suppliers.

APPROVAL OF ADVERTISING CLAIMS FOR TECHNICAL ACCURACY

R&D should have responsibility for the approval of technical accuracy of proposed advertising claims. This is essential to avoid claims being made in the advertising of company products which are indefensible should their scientific accuracy be questioned by FTC.

SAFETY OF COMPANY PRODUCTS

R&D should have the responsibility for procuring information to establish the safety of all company products under U.S. FDA and U.S. Dept. of Agr. requirements. They should also approve the safety of all existing and new products.

CONSUMER SURVEY MATERIALS

Considerable amounts of money are spent for consumer surveys of current and new company products. R&D should have responsibility for establishing that both competitive and company products used for each survey are truly representative of actual production. This means that knowledge and skills built up in R&D through the regular examination of competitors' products will be used to establish that the products surveyed are neither better nor worse than the median quality for the specific products as seen at the store level.

Careful selection is a primary and important step to insure that trustworthy results are obtained in consumer surveys.

SAMPLE FILES AND ANALYSES OF CONSUMER SURVEY SAMPLES

R&D should keep under adequate storage conditions file samples of all consumer survey material at least until the results of the survey have been obtained and digested by the various divisions interested. Research should also, during the course of the consumer survey, accumulate any analytical information which they deem to be helpful for interpretation of the survey data. This should be made available to the market research head for inclusion in the final survey report.

ACCUMULATION OF BACKGROUND INFORMATION ON PRODUCTS OF COMPANIES BEING CONSIDERED FOR ACQUISITION

From time to time various concerns may be considered for acquisition. After preliminary consideration company management may wish to have R&D survey and appraise products which the particular company has on the market. Such appraisals may involve the determination of chemical composition, the evaluation of selected products against the company's similar products (if such are marketed), and (if not), against the largest selling competitive items. The latter may take the form of RGP blind tests.

While such information may seem a bit extensive, it may prove very useful if the acquisition is made, since it furnishes a sound basis for early company action in the area of product improvement and planning for new products to be added to the existing line.

ASSISTANCE IN SELECTION OF CONSULTANTS FOR SPECIAL PROBLEMS

Most companies in the mass marketing field often need answers to problems which cannot be handled in their own Research Divisions either because of lack of special facilities or researchers trained in the required field. For example, in establishing the safety of a food additive, most concerns find it expedient to retain a consulting laboratory to carry out the necessary animal feeding tests with several types of animals as required by the U.S. FDA. Still another type of problem for which consultants are useful is the compilation of information in connection with proposed advertising claims. Such data from a reputable consultant will generally carry more weight with a federal agency, such as the FTC, than will data from a company's own R&D Division.

R&D should have the responsibility for recommending the consultants to be employed for various projects in the research and development area, as they arise. It should also keep in touch with the work being done by the consultant and report and evaluate progress to management.

It is essential, therefore, in the carrying out of this function, that the R&D head be informed about available consulting groups which might adequately handle work in specific fields. Personal acquaintance with the heads of the various consulting firms is helpful as is also some knowledge about their equipment and the competence of their staffs.

COMMUNICATIONS WITH VARIOUS DIVISIONS OF THE COMPANY

Communications are essential for the orderly and efficient operation of any company. Since R&D impinges on the operations of many segments of the company, a system for regular communications is necessary.

Marketing, market research, and production are, perhaps, the most important groups with whom communications should be maintained on a regular basis. This is facilitated by monthly meetings between each of the above groups and R&D. These should be arranged at some mutually-agreed-upon regular date and interested personnel including the R&D head and the head of the group participating should attend. For example, at the marketing-R&D meeting, the marketing head and selected brand managers should attend as well as the research director, selected product development people, and possibly some process development people.

Generally, such meetings should have an R&D department head as chairman, an agenda should be prepared in advance listing problems to be discussed, and each group should have someone present who is prepared to discuss each of the subjects included in the agenda by that group.

Brief reports on each item and any decisions made or action taken should be included in minutes of the meeting. These minutes should also contain a list of those present from each group.

The minutes should be circulated to all present and to selected senior management, including the chief executive officer.

A file of the minutes of each R&D communications meeting is helpful, if subsequently a determination of action and decisions previously taken is needed. Reference to the minutes will frequently save time and prevent misunderstandings.

INTRADIVISIONAL COMMUNICATIONS FOR THE R&D DIVISION

In the R&D division of a mass marketing company, there exists a greater need for regular communications within the division than in most other types of companies. This is because the programs are so closely integrated with marketing needs which can only be successfully met through effective coordination of research efforts.

The average research man does not have an appreciation of the urgency of marketing requirements. He usually feels that his own ideas are more valuable than the projects which may have been agreed upon by other divisions of the company as representing the desired goals and priorities for the program. This presents both an efficiency and morale problem for R&D management.

The program of interdivisional communications recommended earlier, serves to some extent to inculcate, in a few of the higher research managerial people, an appreciation of the goals the company desires to reach and to infuse in them a sense of urgency which is one of the requisites for accomplishment. Intradivisional meetings, if effectively managed, will diffuse some of the information developed in the interdivisional meetings and thereby contribute significantly to all research workers' understanding of the company's research needs and likewise promote improvement in morale and stimulate enthusiasm.

Some of the desirable purposes which may be accomplished by means of intradivisional meetings are:

(1) An opportunity for research people actually doing the work to present to R&D management brief and accurate reports of progress on assigned projects.

(2) Suggestions may be given to the reporter about possible solutions of problems which he may have encountered, especially those in which a fresh point of view may be helpful. These may come from anyone present, but will frequently come from researchers in the supporting science groups.

(3) Attendance by supporting scientists leads them to visualize areas where their specialty may be useful in advancing accepted projects, thus promoting morale and unity in the division.

(4) The meetings offer a convenient and suitable opportunity for R&D management to pass on to the staff such information as is appropriate about reasons for dropping projects or changing of emphasis. This is quite necessary from the morale standpoint because researchers need to have some sound reason for program changes otherwise they tend to attribute them to the whims of management and not to revised planning necessitated by changes in the marketing situation.

Secretaries of each of the meetings should record briefly the topics discussed, progress made, and proposed next steps. These should be prepared and distributed within 2 or 3 days after the meeting to all research managerial personnel, to people attending the meeting, and to any other individuals designated by the R&D head.

The meetings should be open to any member of the R&D staff who is interested in attending. Members of the supporting science and other supporting departments should be encouraged to attend for what they may be able to contribute to the discussion and the ideas which they may generate.

Because of the variability in size and organization of research staffs of various concerns, no specific recommendations can be made about the number and frequency of intradivisional meetings. At least two, a Product Development and a Process Development Meeting, should be instituted at regular monthly intervals.

INTERDIVISIONAL COMMUNICATIONS MECHANISM FOR THE R&D DIVISION

This section is concerned with the methodology by which the operation of the division can be improved and the expeditious advancement of the agreed-upon projects to the desired goals can be facilitated. As a preliminary approach, it may be desirable to set forth the philosophy underlying the recommendations for the several committees and meetings, and how these relate to the company's R&D program.

The use of the term "Committee" has been strictly limited because all too frequently it becomes a device to dilute responsibility and thus accomplishes very little in the way of worthwhile recommendations in relation to the time of the members consumed in the deliberations. The group meetings are set up primarily for communication, discussion, exchange of information, and agreements on action to be taken. There is a distinct need in every mass marketing organization for meetings whose function is to promote interdivisional communications between R&D and other divisions. Reviews of the status of projects of mutual interest, questions, answers, discussion, and agreements on future action are, in general, proper subject matter for these meetings. They are an effective time-saving mechanism because at a single time and place all matters proper for the meeting can be cleared up, thereby eliminating much letter writing, telephone time, and even interplant visits.

Each meeting should be based on a predetermined but brief agenda, usually no more than a word or phrase to cover each subject. The preparation and circulation of a succinct report on the subjects discussed, with specific mention of each action agreed upon, covering both the nature, the timing, as well as the persons, departments, and divisions responsible serves both as a record and an important means of avoiding future controversies.

The development of new product projects and of projects aimed toward the improvement of currently marketed products to equal or surpass competitive lines may be effectively handled by a product projects meeting. The marketing division is responsible for the advertising and sale of both new and improved products. It should make the primary recommendations as to the kind of merchandise it can sell. It is not an easy assignment; but if properly carried out, it will go a long way to ensure that research expenditures are not wasted in developing products which the marketing division has no interest in commercializing. In other words, a systematic study of the probable potentials of a product recommended for research development should be made and a reasonably clear conceptual image of the kind of product desired should be set forth as a guide for development.

It should be pointed out that development of a product conceptualization to reality with all the "must haves" may prove to be impossible within a reasonable time; so provision should be made for regular communications with marketing to consider compromises which may be acceptable. For this and other purposes, meetings with marketing people should be scheduled on a monthly basis.

The R&D head and other research personnel, as desired, should attend. Marketing should be represented by the marketing head, the appropriate marketing managers, and such others as they may invite.

As noted earlier, agendas should be agreed upon in advance and selected representatives should be prepared to report on particular items from the marketing or research point of view. The reports and discussions serve to keep each group aware of the others' attitudes and thinking and save much time in correspondence and may prevent costly errors because of misunderstandings. These meetings also have important educational function since R&D becomes thoroughly aware of marketing problems and marketing develops an understanding of R&D problems and attitudes. This mutual understanding of each others' problems is an item of major importance in resultful research and marketing effort.

Processing problems continually arise in every company and often need the joint efforts of production and R&D for their solution. This means that production and research should work closely and keep each other informed of developments in matters of common interest. A regular scheduled monthly meeting between research and production people provides for a regular review of the status of projects, permits questions to be raised and discussed, and agreements to be reached as to scheduling of work to be undertaken by each division. Experience will soon indicate that much correspondence can be eliminated and joint programs facilitated by a regular production-R&D meeting.

Several of the R&D functions described in the earlier report involve cooperation with market research. Among these are: to insure that competitive and company samples for consumer surveys are representative of those in trade channels, to accumulate necessary analytical data to aid in interpreting survey results, and to maintain adequate samples of all consumer survey material. A monthly R&D-marketing research meeting is recommended to answer questions from each group, to agree upon scheduling of projected consumer surveys and associated problems, and discussions of any other matters of joint interest.

The setting up of adequate communications to handle packaging problems in an orderly and prompt manner needs careful consideration. Packaging problems are of major importance to the marketing division. R&D and production have limited but important responsibilities, the former for establishing adequacy of the proposed package for protecting the product and the latter to determine whether it can be packed in existing equipment without undue breakage and excessive down-time. The importance of scheduling and communications to insure that the activities of each division concerned are going forward according to program merits recognition by the development of a company policy for the orderly handling of packaging problems. In this area marketing, purchasing, production, R&D, and legal divisions generally have definite responsibilities for portions of the overall effort in marketing a new product or in placing an existing product on the market in a new type package.

The above discussion clarifies the need for a company policy covering packaging development which adequately sets forth the responsibilities of each division or department and makes provision for adequate communications either through a packaging coordinator or by means of regular meetings of representatives of the groups primarily concerned.

Based on the preceding discussion, the average mass marketing company should consider establishing a suitable number of interdivisional communications meetings. The number will vary somewhat with the needs of the particular company. The following four are generally sufficient to fulfill the requirements of most concerns: (1) Corporate Research Committee (monthly), (2) Marketing-R&D Meeting (monthly), (3) Production-R&D Meeting (monthly), (4) Market Research-R&D Meeting (monthly).

Corporate Research Committee

The Corporate Research Committee is perhaps the most important since its function is to keep top management informed of R&D progress and to insure that the problems and projects are in line with management's plans for the future development of the enterprise. The purposes of this committee are:

(1) Review R&D progress on important projects.

(2) Consider and advise on recommendations by R&D for changes in program or projects resulting from development work.

(3) Give final approval and priority assignments to new product projects and maintenance of current product projects.

(4) Review product and processing difficulties as they may affect the research program.

(5) Consider and agree upon changes in the R&D program necessitated by unexpected developments.

The membership of the Corporate Research Committee will vary depending upon the management structure of the company. It should include the chief executive officer, the marketing head, the production head, and the R&D head. Others may be included if desired. Meetings should preferably be scheduled at a regular semimonthly or monthly period depending upon the size of the research program.

Marketing-R&D Meeting

A regularly scheduled monthly meeting between the marketing and R&D divisions is useful in providing a mechanism for keeping each group informed about the developments in the other as related to their mutual areas of interest. The purposes of such a meeting are:

(1) To receive reports on R&D progress on projects of interest to marketing.

(2) To discuss any questions marketing or R&D wish to bring up about product projects either accepted or under consideration.

(3) To review or introduce for opinions or advice any matters pertaining to the product program.

The composition of this meeting may be variable depending on the size and structure of the company. The R&D head and selected research staff should attend as well as the marketing head and selected people from his staff.

Production-R&D Meeting

A regular monthly production-R&D meeting is helpful in facilitating the operation of each. The purposes of such a meeting are:

(1) To review status of product and process projects of joint interest.

(2) To discuss processing problems which require R&D assistance.

(3) To forecast project developments which may require production action.

(4) To agree upon timing of actions by R&D and production on programs of joint responsibility.

(5) To communicate results of search for causes of "off quality" products and to discuss remedial action.

(6) To review quality control problems and to discuss results of regular inspections of company products purchases at the store level.

(7) To report on any subjects of mutual interest within the responsibilities of each group represented.

The production head and selected members of his staff, the R&D head with selected people from his staff, and a representative from the purchasing section should attend.

Market Research-R&D Meeting

A regular monthly marketing research-R&D division meeting will facilitate the work of each. The purposes of such meetings are:

(1) To discuss proposed consumer surveys and to agree on standards of quality for both company and competitive products to be surveyed.

(2) To determine minimum quality inspection and analytical information required for MRD purposes.

(3) To answer any questions which may arise in MRD and R&D fields of interest.

(4) To coordinate and reach agreements in the handling and interpretation of RGP results.

(5) To discuss other areas of mutual interest to the members.

R&D COMMUNICATIONS WITH SCIENCE COMMUNITY

Contacts with other Scientists

The R&D staff, particularly those of higher scientific attainments, should be encouraged to maintain and expand their contacts with university and industrial scientists. This is one way to avoid the ingrown attitude that leads to a relatively nonproductive and pedestrian staff.

Membership in various scientific societies should be encouraged, as should activity in the local sections of any of these which may be in the vicinity. Some concerns pay annual dues in certain societies for members of their staffs. It is doubtful that such a practice should be recommended, because any scientist should want to belong to groups of his choice and receive the publications at home where he can spend as much time as he wishes to keep up with new developments in his field.

For 1 or 2 R&D people, a company membership in the Industrial Research Institute is desirable. At the semiannual meetings, these people have an opportunity to meet with the leaders in industrial research and to participate in discussions of a very wide variety of research management problems.

R&D should also maintain outside contacts, particularly with the research divisions of suppliers of ingredients, packaging materials, and other items. Good relations of this type are of benefit to concerns since the supplier often finds it possible to pass on results of internal research that may be helpful, and will frequently undertake work on special problems in the hope of expanding his sales.

SUMMARY

The preceding is a brief presentation of generally assigned responsibilities of a Research and Development Division in a mass marketing company. The suggestions made should be regarded as guide lines pointing to ways in which a sense of urgency can be imparted and fostered in an industrial research organization of the type under discussion. Unless such a sense is developed and maintained by proper two-way communications, the group will most certainly have a performance record below the level of their capabilities.

More will be said on the subject of communications in the chapters to follow. The guide presented in this chapter will be followed in the next chapter, which deals with bringing the new product into production.

BIBLIOGRAPHY

Anon. 1970. Leadership, power relationships and decision making. Harvard Business Rev. Reprint Serv., Soldier's Field, Boston.

Anon. 1970. Organizational Development. What makes for strong, flexible, growing organizations. Harvard Business Rev. Reprint Serv., Soldier's Field, Boston.

Anon. 1970. Personnel Management. Harvard Business Rev. Reprint Serv., Soldier's Field, Boston.

Anon. 1970. R&D Management. Harvard Business Rev. Reprint Serv., Soldier's Field, Boston.

Bass, L. W. 1967. The planning of innovation. Chem. Ind. 1671-1675, Oct. 7.

Cotton, R. H. 1970. Management of product development in a large diversified corporation. Food Technol. *24,* No. 4, 36-38.

Fredrickson, E. B., and Lawson, E. W. 1970. The role of technology transfer in product development and marketing.. Res. Management *13,* No. 4, 265-272.

Hilton, P. 1967. New Product Development. Prentice-Hall, Englewood Cliffs, N.J.

Parsons, L. B. 1963. Personal Communication. Ridgewood, N.J.

Sly, W. H. 1968. What is involved in the management of R&D. Food Technol. *23,* No. 1, 41-46.
Wilson, E. B. 1952. An Introduction to Scientific Research. McGraw Hill Book Co., New York.

CHAPTER 7

The Physical New Product

Once we have the viable concept for a new product, and the resources to do what is necessary, the next step is to bring physical form to the idea. The concept must be built into a physical entity. This takes place in laboratory, pilot plant, and factory, and eventually involves nearly all divisions of a company. How the physical form is brought about is the subject of this chapter.

CONSUMER GUIDANCE FOR RESEARCH LABORATORY

In general, once a new product project has been established and assigned to a development technologist, it becomes of utmost importance that he focus his attention on the product concept under study and that the decisions he makes concerning his progress in developing the product become as objective as possible. It is at this point that product development research guidance in the form of organoleptic panels of various types, sizes, and compositions becomes available. For new product development purposes, another type of panel is needed industrially, beyond laboratory taste panels, called generally the Research Guiding Panel (RGP) by Parsons (1963).

In order to put the RGP system into operation, many laboratories require three types of organoleptic testing panels preliminary to the RGP. A typical taste panel arrangement is shown in Fig. 7.1.

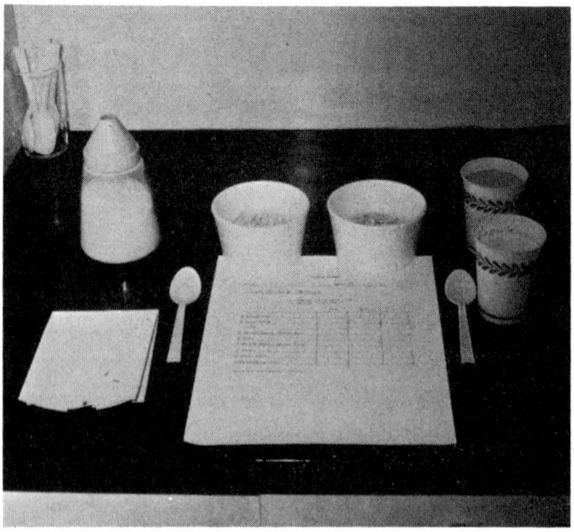

FIG. 7.1. TYPICAL TASTE PANEL ARRANGEMENT

Taste panels are decision-making bodies.

Expert Profile Panels (an Analytical Tool)

Several highly trained experts are organized to identify and establish intensity values for sensory characteristics of products, e.g., what is the major flavor note in sample X? (The sample might have either a mint-chocolate or a chocolate-mint flavor.)

Primary Sensory Panels (a Product Development Tool)

Groups of product development personnel organized into panels evaluate their products at each step in development. The purpose is to yield the best product possible.

Secondary Sensory Panels (a Decision Tool for Measuring Differences)

Samples are tested for significance from control samples (see Fig. 7.2). If the product meets design criteria, it is ready to move into the pilot plant. At this point, a new type of testing panel becomes needed.

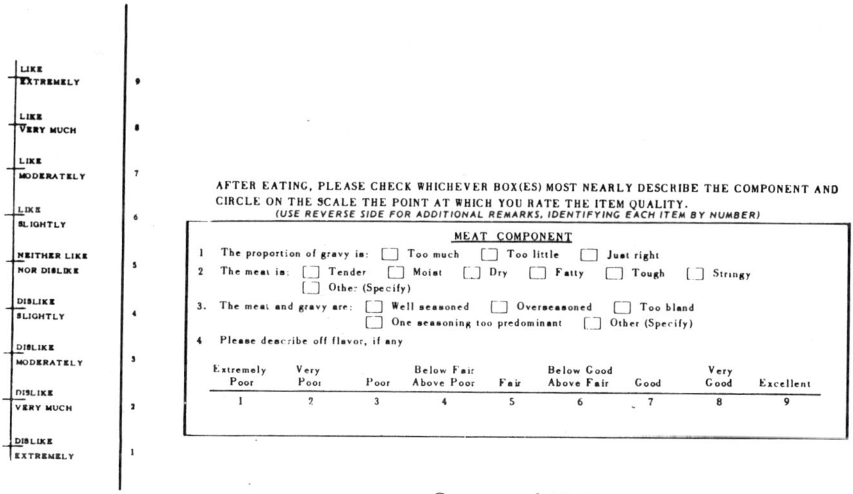

Courtesy of QM Food and Container Institute

FIG. 7.2. SCORE SHEETS USED IN EVALUATION OF FOOD SAMPLES

Food acceptance score sheet (left) and food quality evaluation sheet (right) are examples of those used in food research.

RESEARCH GUIDANCE PANELS

Purpose

RGP represent a relatively inexpensive mechanism for measuring consumer acceptance of products under development by a research laboratory. Information of this type is necessary at a fairly early stage in product development, and is useful in the continuing modifications of properties designed to secure the maximum consumer acceptability within the predetermined cost. Products so developed will generally meet all the criteria demanded in a broader sense.

Panel Organization

RGP's can readily be set up, especially in a metropolitan area where a broad spectrum of people of various racial origins and foreign and native derivation exists, according to Parsons (1963). Voluntary panels of suitable size are organized. Test samples are prepared and distributed (see Fig. 7.3).

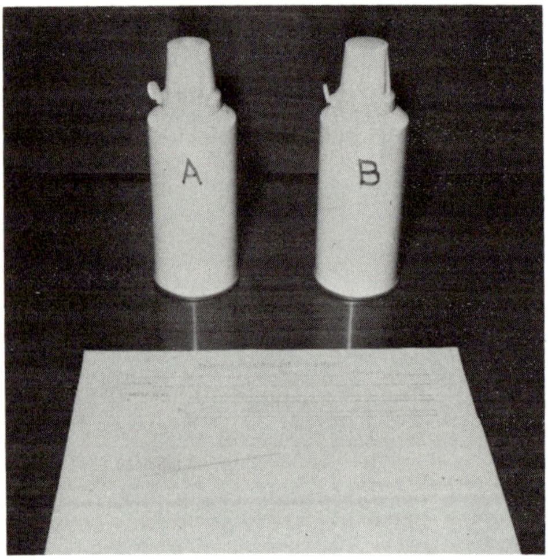

FIG. 7.3. RESEARCH GUIDANCE PANEL (RGP) TEST SAMPLES

Care is needed in sample preparation to prevent comparisons among panel members which might influence results.

Utility of Results

Parsons found excellent agreement between RGP testing and broad scale consumer surveys with wide geographical distribution, using the same paired samples but with much more elaborate questionnaires.

One important point is the early availability of results—weeks as compared to months for the average consumer survey. Another is the cost. An example of the type of result received from a product is shown in Table 7.1.

The above tests are functional and relevant. While results are dependent on a particular product (e.g., ice cream versus hot cereal) and the tasting group dynamics, in general, values on such tests ranging above 7 on a 9 point scale will be potential candidates for nationally distributed products.

TABLE 7.1

EXAMPLE OF RGP RESULTS: MONADIC TEST ON FROZEN STRAWBERRY PIE, SUCH AS MIGHT BE OBTAINED

	Adults	Children	Combined
	Flavor of Filling (%)		
Too strong	47.6	33.0	38.6
Just right	39.7	52.8	47.7
Too low	12.7	14.2	13.7
	Sweetness of Filling (%)		
Too sweet	73.0	47.8	57.5
Just right	27.0	49.8	41.0
Not sweet enough		2.4	1.5
	Number of Flavor Remarks		
Good/like crust	46	Dislike filling	31
Good filling	17	Too sweet	16
Good/tasty/excellent	12	Filling too sweet	15
Liked	10	Dislike	8
Real strawberry flavor	9	Artificial strawberry filling	
Just sweet enough	3	Flavor too strong	5
	Texture Remarks		
Crisp	6	Dry crust	4
Flaky crust	5	Crumbly	3

TYPES OF NEW PRODUCTS FOR A FOOD COMPANY

Technologically, the development of new products for a company usually falls into 1 of 4 categories, in decreasing order of difficulty:

(1) New factory, new technology and equipment (i.e., new raw materials, new processing equipment, new packaging, warehousing, and distribution needs, etc.).

(2) Existing factory, new types of technology and equipment (i.e., new process and/or new packaging equipment, etc.).

(3) Existing factory, some minor new technology and equipment (i.e., new step in process, or new packaging equipment, etc.).

(4) Existing factory, new entry using existing technology and equipment (i.e., new canned food, new frozen food, new dry mix, new cookie, etc.).

CATEGORY I–COMPLETELY NEW PRODUCT

Comprehensive market analysis is often undertaken as the initial phase of a program. Next, research laboratory personnel in conjunction with market research specialists develop new product ideas and concepts embracing potential consumer interests and needs revealed by the market analysis. The most meaningful and fruitful are technological innovations which would lead to a whole family of additional products.

Concept Testing

These concepts are then screened for technical feasibility, developed to prototype stages, and consumer researched to establish prototype acceptability. The decision, more often than not, revolves around the compatibility of the new product with long-range company plans, and the choices available for commitment of available resources.

The following outline gives the general sequence of events in more detail for a formal development program. Close coordination is required at each stage with all operating elements involved.

Prototype Product

This type of product development deals with an entirely new product concept where no known product on the market meets the anticipated need. A prototype product must be developed, and the optimum method of preservation identified. (A typical product development laboratory is shown in Fig. 7.4.)

Courtesy of Nabisco

FIG. 7.4. TYPICAL PRODUCT DEVELOPMENT AREA; NOTE ABSENCE OF LARGE EQUIPMENT

Process Development

When a new product reaches the development stage where it appears feasible, the laboratory begins to plan an economical process for its industrial manufacture. (A scaled-up production unit is shown in Fig. 7.5.) In process development many factors must receive consideration, such as: (1) Can it be made in an existing factory? (2) Should it be a continuous, semicontinuous, or batch process? (See Fig. 7.6.) (3) What physical properties of the ingredients, mixes pre-

Courtesy of American Machine and Foundry Co.

FIG. 7.5. CONTINUOUS BREADMAKING SYSTEM SHOWING COMPONENTS FOR AUTOMATIC DOUGH MIXING, EXTRUDING, DIVIDING, AND PANNING

FIG. 7.6. AN EXAMPLE OF CONTINUOUS FLOW EQUIPMENT

Product can be preheated, sterilized at high temperatures and pressures, then cooled for discharge at atmospheric pressure.

pared during formulation, and the product itself are needed for process equipment design? Such matters as temperature-viscosity relations, heat transfer coefficients, emulsion stabilities, density range, flow rates, and specific heat data are among the factors frequently requiring study.

A more complete check-list of the factors requiring specific attention are presented later in this chapter. The specialized knowledge of production personnel may also be called upon at this stage, as required. However, for our present purpose, the following sequence generally occurs: (1) Establish tentative process specifications. (2) Organize necessary pilot plant facilities. (3) Produce pilot plant products and submit to taste panels and run RGP. (4) Coordinate with marketing unit. (5) Produce sufficient pilot plant quantities of products for consumer research. (6) Coordinate with production, purchasing and legal units. (7) Formalize product and process specifications by preparing tentative (a) raw material and packaging specifications, (b) process specifications, (c) quality control manuals, (d) product and process operating manuals, and (e) patent disclosures covering inventions.

Public Health Clearance

Complete studies to establish (1) absence of public health hazards in all raw and finished products; and (2) obtain laboratory microbiological clearance.

Packaged Product Storage Studies

Establish storage life of packaged products by (1) actual tests under several temperature and humidity conditions, taking careful note of the information presented in Chap. 13; (2) insure package integrity by abuse tests; and (3) perform needed organoleptic tests.

Finalize Specifications

Correct any defects in packaged products (1) revealed by the above studies; (2) complete all specifications; and (3) submit specifications for establishment of costs.

Develop Advertising Claims

In conjunction with marketing unit (1) develop statements concerning desirable characteristics and uses of developed products for potential uses in advertising; and (2) provide data to support advertising claims.

Preproduction Runs

In many companies, the production organization has an engineering group for design and maintenance of plant.

As an example, research will generally accumulate all the data on physical properties required for process equipment selection or design. Production engineering can then lay out the designed process with due regard to available space, available equipment, and other production considerations.

Research will generally have worked with small scale pilot equipment to obtain the necessary physical information. Presumably also, the pilot plant will have used either specially built or available commercial equipment in producing pilot lots of the proposed product. It is well to keep in mind that small pilot operations frequently give differing results from plant operations, where there may be from 10- to 100-fold increase in size.

Market Research

Provide Products for Field Testing.—All necessary information concerning product acceptance based on actual home-use tests must be completed successfully. Here is where the idea of how many times a week the product will be used (which gives some indication of the production volumes to anticipate), what the consumer feels the price should be, and her reactions to the product overall are unfolded. This process of finalization is undertaken with potential consumers in test market areas. Typical results of market research are shown in Table 7.2.

TABLE 7.2

TYPICAL CONCEPT TEST SUMMARY

	Prototype—Child Food Product			
	A N = 188 (%)	B N = 188 (%)	C N = 186 (%)	D N = 186 (%)
Definite purchase intent	34	44	32	31
Total purchase intent	72	77	72	65
Inclined to switch	33	32	38	33
Preference				
Housewives	38	52	46	45
Children	28	55	29	53
Product Ratings[1]				
Overall appeal	7.3	7.7	7.1	7.0
Natural rather than artificial taste	7.9	7.7	7.9	7.4
Retaining freshness	8.9	9.0	9.0	9.1
Having an attractive color	8.1	8.5	8.1	8.4
Having a pleasant texture	8.1	8.1	7.9	7.8
Having an attractive appearance	7.0	9.0	7.2	8.9
Best overall taste	7.3	7.5	7.6	6.9
Having an appealing shape	7.0	9.0	7.1	8.7

1 1-10 rating, with 10 being highest possible score.

Timing

When existing production and packaging facilities are available, the production of test market quantities may be anticipated in less than 50 weeks from the time of initiation of a study.

New proposals requiring new facilities and equipment may take from 2 to 4 yr to reach the market place regionally. It will take approximately 1 yr from this point to complete regional test market studies prior to launching nationally.

Overall, it generally requires from 3 to 5 yr to bring a completely new product into national distribution. It may take longer.

CATEGORY II—NEW PRODUCT FOR COMPANY-EXISTING COMPETITOR PRODUCT

The Idea Sources

First of all, the ideas for the improved products must be uncovered. Experience indicates that ideas frequently come from unexpected sources, as noted in previous chapters. All should receive attention with due regard to the legal aspects of company liability, particularly in the case of submissions from outside a company, as will be noted in a later chapter.

"Must Have"–"Would Like" Specifications

Once a novel product idea has passed marketing and research scrutiny, the next step is for the marketing unit to set up specifications for the product they would like to market. This also takes the form of a description of properties it should have from their point of view. The preliminary specifications should embrace the "must haves" and "would likes." Now the immediate goal is to develop a prototype product and to learn by means of suitable testing what the consumer reaction may be. (See section on RGP.) Here the simplest problems concern a product which is designed to fulfill a given need better than any product currently on the market. RGP blind testing of the prototype versus the leading product will serve to establish consumer likes, dislikes, and preference patterns.

CATEGORIES III AND IV

New products for a company falling into these categories are undertaken with less difficulty. While all the steps and testings are performed, these products are usually based on minor technological changes. They can be brought about with more dispatch and less involvement of personnel and resources, since the area has been successfully explored before. Therefore, the risks are reduced in most stages of development.

INDUSTRIAL APPLICATION OF NEW TECHNOLOGY

While science and technology are thought of as the relation of things and processes to each other, when attempts are made to relate them to man, another component is added—common sense. The application of science and technology eventually involves people, and their harmonious and fruitful interrelationships. Some examples follow.

Information Required Prior to Launching a New Product

Some feeling for the myriad details and decisions required prior to reaching the market place with a new product can be obtained from the following generalized listings:

The Product.—Its specifications; formulation tolerance; analytical techniques for specifications; sampling techniques for specification control; patents on the product itself; toxicity clearance; quality control manual; and uniformity of production from panel evaluations.

Product Storage Specifications.—At normal conditions; at accelerated conditions; and correlation between normal and accelerated storage.

Packaging.—Specifications for carton development (type, size, and materials); label development; label declarations; recipe preparation for label, and design of label; case specifications; shipping tests; and field storage and handling of packaged product.

Marketing.—Consumer acceptance of the product through home-use tests and market tests; consumer acceptance of product versus nearest competition; determination of quality/price specification-consumer desires; determination of consumer habits in use of product (i.e., full package, part of package, half package, etc.); what is the best size package; time to start filling distribution channels; advertising and when to initiate; trade markets; identification of potential competing products; study of market demand; volume expected and location of market demand; legal requirements for its sale (use of patents, of materials involved, and other requirements); and seasonability of market.

Raw Materials Specifications.—For each ingredient: size, purity, composition, contaminants, variety, color, and others and the variations permitted in each specification; analytical procedures for identifying specifications; sampling procedures; toxicity clearance; availability of the materials meeting specifications (location—geographic and suppliers; time—seasonal or periodically variable supply within seasons or periods, variable supply from season to season or period to period, and others); characteristics of material available (technical or chemical, physical, varietal, and others); list of sources; adequacy of supply; market influence on supply and resulting fluctuations; costs (fluctuations, compositionwise, and qualitywise); transportation; government; market speculation influences; selection of material for processing (techniques needed to establish whether material from an aggregate contains some "out of" specification material, and the possible uses of substitute specification material); handling (equipment needed, techniques required, requirements during transportation, personnel training required, and hazards involved); aid required to suppliers—either industrial or agricultural; legal problems; tax, tariff, and patent problems; purchasing (contractual agreements, special training or knowledge required, and the amount of effort required).

Raw Material Storage.—Containers and conditions required (normal or special); effect of storage conditions on quality and quantity; and facilities investigation.

Process Development.—Process conditions; research, pilot plant, production, optimum conditions for processing, and tolerance of processing conditions; in-process specifications; peculiarities of the process; balance between quality and possible cost reduction; raw material effects on process and product; uniformity of in-process control; process formulations, basic formula development, tolerance limits of formula, and possible substitutions or variations of formula; process description; adequate presentation of information; subsequent modifications; yield-sources; causes, importance, and methods of reducing losses; personnel requirements (number, training, job descriptions, points of high usage, and methods of reducing the number of people in time); utility requirements (quantity, both normal and maximum demand, quality, and possibility of reducing them); process sanitation; process safety and problems; patents (patent infringement of other processes, patents on new process); licenses required; waste disposal (amounts of waste, composition of wastes, possible combinations of waste for treatment, and possible treatments); time-motion studies (yield versus labor, quality versus yield, and others); equipment performance (capacity, process problems, ease of operating, stability, and maintenance problems); development of quality control techniques, and quality control manual; capital costs; process costs; and process requirements for special building facilities.

Facilities.—Definition of facilities including flowsheet and specifications, equipment, utilities, labor, area requirements, building, land, costs, and design; construction; delivery times for equipment; and installation schedule.

Plant Location.—General area of plant location (specific area, location within a general geographical locality, location within a given plant); land characteristics (virgin or filled land, cavernous, rocky, sandy, or other); utilities available in the desired area (power, water, telephone, gas, and others); suitability and adequacy of water supply (for processing, for power house, and others, and treatment needed); availability of required fuel; labor survey (availability, qualified, labor laws, and other data); living conditions; effect of processed plant on nearby industries; effect of climate on process; waste disposal (legal requirements and fitting process requirements to legal requirements); air pollution requirements of community; transportation availability; taxes and other fees; legal requirements; relation between raw material supply and the location of markets; problems peculiar to the plant construction at that locality; and acceptability by the local people.

Costs.—Materials (process raw materials, packaging materials, auxiliary materials, and cleaning); processing (direct labor, indirect costs, supervision, and marketing; other costs; and overhead.

Plant Operation.—Job descriptions; acquiring personnel; personnel training; plant start-up; operating manual preparation (initial and revisions); sanitation check; check on safety and health conditions; arrangements for materials for production (raw materials, packaging materials, and miscellaneous materials).

Normal Production.—Periodic review of market performance (objectives are being met); and maintenance (periodic up-dating of product/process to yield "new improved" product).

It is obvious that the above noted matters need not all become the subject of new study for a company, and there also could be other items added to these lists. However, it also should be obvious that an organization involving a host of professional personnel is needed.

INTERLOCKING ACTIVITIES OF PEOPLE AND ORGANIZATIONS

A further insight into the division of responsibility, efforts, and the communications needed between organizations in bringing out a new product may be found in just one aspect of it—establishing the package for a new product. In large food manufacturing companies the following pattern is often found:

Marketing Unit.—Decision on the type of package, cost, design, labeling, and other information for casing of shelf units.

Purchasing Unit.—Work with suppliers to obtain submissions of new or modified types of packaging for marketing unit inspection and for preliminary consideration by the research laboratory and production unit.

Production Unit.—Preliminary advice on the feasibility of proposed packaging changes as related to operation of packaging equipment and final testing of operating characteristics on available equipment in the event of acceptance by the marketing unit.

Laboratory.—The preliminary and final determination of the adequacy of the proposed package to protect the product. It should also assist marketing and purchasing units in obtaining preliminary samples of packaging materials. This assistance should be on request of these two groups and generally will take the form of discussions with suppliers' technical people on needs and problems as they arise.

Legal Unit.—Generally, the legal unit has responsibility to insure that labeling meets the requirements of U.S. FDA, FTC, and meat inspection service or other government agencies that may be involved in the United States. (In other countries, similar governmental agencies frequently become involved.)

The effectiveness of the R&D team will be revealed in the dispatch with which they bring out new products which meet established goals. Industrial research laboratories generally devote a major portion of their resources and personnel to new and improved product and process projects. The flow of knowledge to sustain these activities originates from the scientific and technical laboratory people, working in a laboratory unit such as that shown in Fig. 7.7.

COMPETITIVE ACTIONS

Another factor to receive serious consideration is that inventions, being what they are, often can occur to more than one inventing team or person in more than one company at nearly the same time. It must be assumed that this will occur and plans drawn accordingly.

Given success in launching a new product, various means are available to measure the progress of the product as competitors and markets grow. One of the widely-used services to keep abreast of market activity and product improvement is the Nielsen Food Index. In a regularly updated review, product and competitor shares of markets, regionally and nationally, are estimated. Based on these and internally-produced information in a company, future plans and needs unfold. The general approach used in such reports is shown in Table 7.3.

Courtesy of Nabisco

FIG. 7.7. A TYPICAL SCIENCE-TECHNOLOGY TYPE LABORATORY IN INDUSTRY

These units are the basic spawning ground for new products for industry.

TABLE 7.3

TYPE OF INFORMATION GENERATED BY MARKET ANALYSIS, CONCERNING PRODUCT MOVEMENT THROUGH RETAIL CHANNELS, ONCE PRODUCT[1] REACHES NATIONAL DISTRIBUTION

	% Total Instant Beverage Sales			
	Product "A"		Product "B"	
U.S. Sales[2] Total	May-June 8.3	July-August 8.2	May-June 8.8	July-August 8.9
	U.S. Sales[3] Regional Shares			
New England	18	17	20	21
Mid-Atlantic	25	24	30	30
East Central	20	20	24	24
West Central	15	15	10	10
Southeast	5	5	7	7
Southwest	4	4	1	1
Pacific	10	10	7	6
Other	3	4	1	1
	U.S. Sales[4] Selected Cities			
New York	11.5	10.8	18.9	18.9
Chicago	16.2	16.9	10.1	9.5
Los Angeles	7.3	6.6	12.5	13.2

1 For purpose of example, assume products are Instant Beverages, and Product "A" is being challenged by Product "B."
2 Share of total market.
3 Distribution of share regionally.
4 Percent of total sold (in selected cities) of new products compared to the total instant beverages sold in those cities.
Note: Other minor competitive products also exist. Analysis focused on two main entries.

SUMMARY

Science and technology are key building blocks of the food industry. In one way or another, all food products draw upon them; the parameters within which successful new products can exist and existing products can be improved are rigidly defined by them.

Science and technology continue to grow and expand, as discussed earlier; one discovery leads to another, unfolding new possibilities. As these occur, they will be physically applied to new products, using the basic approach described in this chapter.

While it is clear how a new product can be brought into being, what happens when there are 10, or 100, or 1000 such projects underway in various stages of development? How these are organized and managed is the subject of the next chapter.

BIBLIOGRAPHY

Amerine, M. A., Pangborn, R. M., and Roessler, E. B. 1965. Principles of Sensory Evaluation of Food. Academic Press, New York.

Anon. 1968. ASTM Committee E-18 on sensory evaluation of materials and products, Am. Soc. for Testing and Materials. Manual Sensory Testing Methods *STP 434,* Am. Soc. Testing Materials, Philadelphia.

Bass, W. L. 1967. The planning of innovation. Chem. Ind. *39,* 1671-1675, Oct. 7.

Bates, B. 1967. Venture group approach to new products. Assoc. Natl. Advertisers Workshop, New York.

Bright, W. M. 1968. Science and mergers and acquisitions in corporate growth. Food Prod. Develop. *2,* No. 4, 43-50.

Calvin, L. D., and Sather, L. A. 1959. A comparison of student preference panels with a household consumer panel. Food Technol. *13,* 469-472.

Carson, G. D. 1969. Product development decisions using Bayesian decision theory. Food Prod. Develop. *2,* No. 6, 46-50.

Caul, J. F., and Taymond, S. A. 1967. Comparison of local and national consumer panels in a paired preference test. I. Statistical data. J. Soc. Cosmetic Chemists *18,* 123-133.

Dethmers, A. E. 1968. Experienced vs. inexperienced judges for preference testing. Food Prod. Develop. *2,* No. 5, 23-25.

Fairbrother, J. G. 1968. Compressed formulated products. Food Technol. *22,* No. 12, 100-103.

Lavidge, R. J. 1968. Equating market opportunities with product development potential. Food Prod. Develop. *2,* No. 5, 28-31.

Miller, P. G., Nair, J. H., and Harriman, A. J. 1955. A household and a laboratory type of panel for testing consumer preference. Food Technol. *9,* 445-449.

Orman, A. D. 1967. A systematic approach to screening new product ideas. Assoc. Natl. Advertisers Workshop on New Products and Services, New York.

Parsons, L. B. 1963. Personal communication. Ridgewood, N.J.

Schwartz, N., and Pratt, C. H. 1956. Simultaneous vs. successive presentation in a paired comparison situation. Food Res. *21,* 103-108.

Simone, M., Leonard, S., Hinreiner, E., and Valdes, R. M. 1956. Consumer studies on sweetness of canned cling peaches. Food Technol. *10,* 279-282.

Standal, B. R., and Kian, K. D. 1968. Nutritive quality of simulated milk mixtures prepared from tropical plant products. Food Technol. *33,* No. 4, 426-431.

CHAPTER 8

New Products and Research Laboratory Management

The food industries of today have found increasing opportunities in successfully producing and marketing wholesome new products which are made increasingly more stable in storage, permitting widespread distribution to growing numbers of consumers in expanding domestic and global markets. One feature the food industries hold in common is that there usually is a research laboratory of one kind or another involved in new product development. How are these organizations put together, what role do they play, how do they go about doing their part, and how do they manage their activities to insure the new product is developed on target? And, how is it managed on a global basis?

These organizations continue to develop a manifold of new products. The success of their efforts is proved in the market place. What goes on?

INTRODUCTION TO INDUSTRIAL PRODUCT DEVELOPMENT

The successful and expeditious development of new products which meet marketing and production requirements is a major function of industrial food research laboratories. This function, however, involves most other elements of a company, and, in fact, to be effective, must truly be a company "team" effort.

Since many research, marketing, production, purchasing, legal, and financial personnel become involved, product development must be a coordinated and controlled company or corporate function. Its effectiveness will be contingent on close working relationships among all members of the team and the smooth integration of their efforts toward desired ends.

First, let us examine how industrial research laboratories are organized and managed to do their work.

TABLE 8.1

TYPICAL RESEARCH LABORATORY MISSION STATEMENT

1. Develop and keep abreast of scientific and technical knowledge in areas of vested corporate interests
2. Apply this technical knowledge by developing new and improved products and processes for the corporation
3. Establish, issue, and maintain standards and specifications for raw materials and finished packaged products bought and/or sold by the corporation
4. Audit adherence to the approved corporate quality standards and recommend corrective action as required
5. Represent the corporation in matters involving laboratory sciences and technologies

ORGANIZATION AND MANAGEMENT OF R&D

The research laboratory is provided the resources to accomplish its missions. An example of such a mission is shown in Table 8.1.

The basic functions in managing a research laboratory fall into headings such as forecasting, planning, organizing, operating, and controlling to accomplish assigned missions. While, as we have seen, these principles apply equally well to the management of other functions and organizations, research management differs significantly in terms of the predictability of results to be achieved and the nature of the people at work.

Research results are difficult to predict since they are concerned with unknowns. The personnel in a research organization usually are of a level of intelligence and educational background which is rarely met in such concentration in other business activities. These considerations influence how R&D is managed.

Forecasting

A forecast of company objectives must be drawn from the mission statements with which research management can formulate R&D plans and policies to serve long-term company needs. The director of the organization next must take into consideration the scientific resources available to him, and set forth a list of individual projects needed to achieve company objectives. The list may contain dozens or hundreds of projects, depending on the company. This leads him to the planning step.

Planning

Research projects needing attention are then evaluated in depth and a detailed work plan emerges, including some system of priorities for individual projects and time frames in which the objectives must be achieved. This leads to the problem of planning individual research projects.

In the planning stage, it is usually necessary that, for each project, forecasts be made of time, costs, and labor requirements. Some laboratories use critical path techniques for this purpose, to be discussed below.

Organization

Research and development laboratories in industry, Bass (1967) found, generally are organized either in terms of scientific disciplines or follow a product orientation. The latter case is characteristic of the smaller organization; the large organizations tend to organize to accommodate the best of both types of orientations (see Fig. 8.1 and 8.2).

WORK ORGANIZATION

In the United States, research personnel are organized in multiples of about 5. As a general practice, this can be expanded at will, to 25, 125, etc. Ordinarily, a manager ought to have not more than six subordinates reporting to him directly. In a large laboratory, organization should be split into a small number of

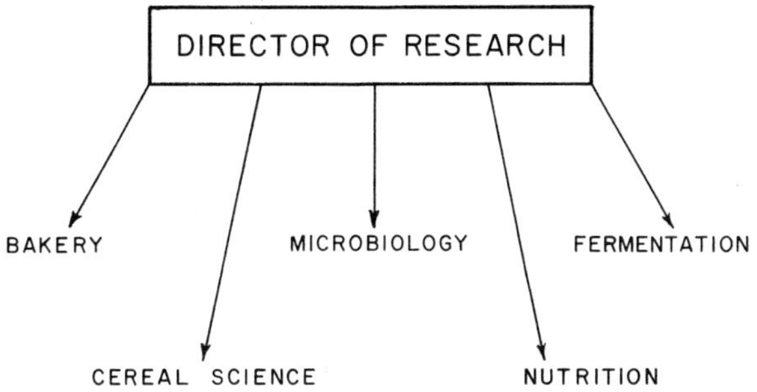

Courtesy of Sly (1968)

FIG. 8.1. ORGANIZATION CHART FOR A DISCIPLINE-ORIENTED CENTER

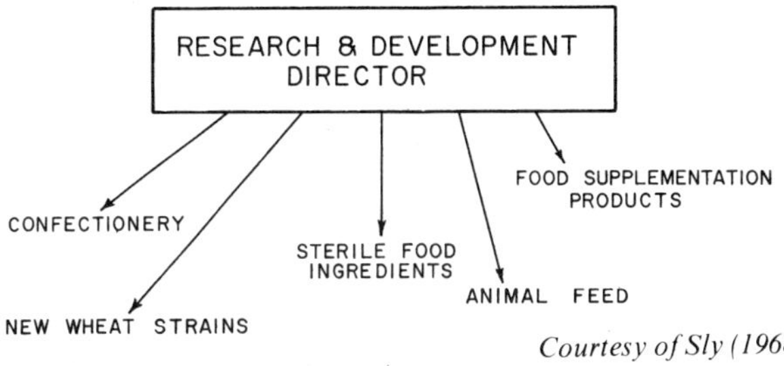

Courtesy of Sly (1968)

FIG. 8.2. ORGANIZATION CHART FOR A PRODUCT-ORIENTED R&D CENTER

large departments or groups of departments, each containing a manageable number of project teams or members of teams. Sly (1968) compared such structures with the organization normally encountered in a production unit in an interesting way: the people at the lowest organization level in the research unit correspond to the factory workers on the production line. In most cases the latter would consist of unskilled labor, whereas in the research laboratory all would

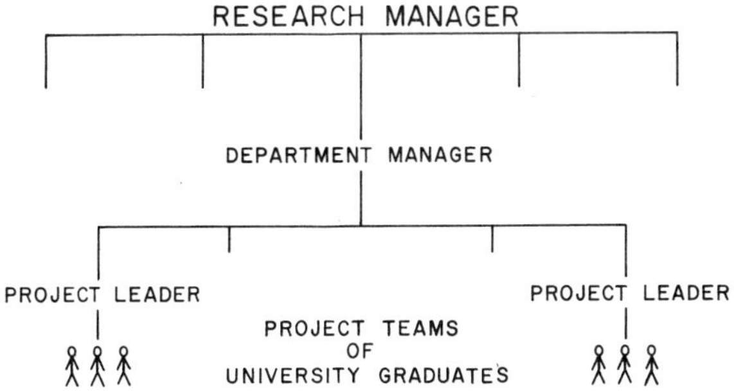

Courtesy of Sly (1968)

FIG. 8.3. ORGANIZATION OF PERSONNEL IN A RESEARCH UNIT TO FACILITATE COMMUNICATIONS AND MANAGEMENT CONTROL have some scientific training and most would be university graduates (Fig. 8.3). The ratio is generally about 1 : 1 overall, of graduate to technician.

This illustrates one of the fundamental differences between research management and the management of other types of business activity.

In particular, the technicians in research are generally highly intelligent people, most of whom have had partial college or university training; nearly all expect to be informed about the underlying policies and objectives which determine the type of work they are asked to do. The communication of this information helps to ensure that all identify with their projects and with the objectives of the company. This contrasts with ordinary factory work. Yet the technique is such that it might be more widely used, the degree depending upon the people involved.

COMMAND FUNCTIONS

The research director is usually able to be in direct contact with all project teams at all times in a small research organization. In the larger units, some command system is called forth to move projects. How does a company, small or large or diversified, develop its command system? Let us explore each type.

Independent Regional Company

Command over the course of events in a regional company usually can be effected with relative ease, since all elements of the organization are generally in one location. Usually, this situation finds a small number of projects as well. For this situation, direct command is possible. It is aided by an effective new product program statement from the management. The essentials listed by Hilton (1967) are:

(1) Proper direction from the company president including a clear policy guideline, counterbalances on major decisions, and constant attention and support to personnel responsible for new product development, company-wide.

(2) Special organizational structures that devote efforts mainly or exclusively to planning and executing new product policies on a day-to-day basis. One individual or a committee should be assigned complete responsibility for the job of handling new product planning. The form may vary from company to company.

(3) Careful reevaluation of each department's contribution to new products. It is easy to see where the research and development laboratory fits. Increasingly, firms are also recognizing the special contributions of sales, production, and other groups in generating new product ideas. Nearly every department has a particular contribution to make—and defining these contributions is one of the most challenging elements in today's new marketing approach.

(4) Written policies and procedures. Written new product policy guidelines communicate clear and firm policy set by top management. Reliance on verbal instructions runs the danger of conflicting interpretations of policy, and excessive reliance on the memory and experience of 1 or 2 individuals.

(5) Intensive review at each stage of new product planning and execution is needed. This includes frequent checks on laboratory progress at each step of the way.

Large National Corporation

How does a large national corporation manage laboratory efforts? How does a large, diversified, global corporation approach the management of laboratory product development? How it is accomplished in one such corporation with a large national food company holding was reviewed by Cotton (1970).

In this case they blend decentralized profit centers and laboratories with overall worldwide technical management. They found that industry, market, and government requirements differ so much country by country or industry by industry that a single central research development and engineering laboratory would make little sense. Nevertheless, a world headquarters technical group takes responsibility for the annual $300,000,000 technical research effort while encouraging maximum self sufficiency in the units. Communications tied to good planning and regular progress reviews make the system operable, Cotton reported.

As with all other units of the parent corporation, the national company provides an annual business plan. This is the chart by which R&D steers its course. Planned growth calls for new products.

Product development projects are developed by the national research department in very close cooperation with top marketing and production executives. Copies of project reports go to the World Headquarters Technical Department. Liaison between that department and the national laboratory is provided for at appropriate management levels.

The national research department looks to the World Headquarters Technical Department for long-range thinking, for provision of technical skills not available "in-house," for aid in monitoring research progress, and for the scouting of new opportunities. The national research laboratory does product development work for 4 business divisions—bread, cake, frozen foods, and salted snacks—which have combined annual sales of more than $700 million. These business divisions are headed by corporate officers headquartered on the same site as the laboratory. Now, how do they manage product development? First, they define objectives in close cooperation with marketing. Then, they put together a team.

Cotton's typical development team includes a production man, a marketing man, and an R&D man. These development trios cut across department lines both in and out of the laboratory. The R&D member obtains aid, as it is needed, from other laboratory groups.

The R&D trio member visits plants frequently for such things as start-ups, trouble shooting, product improvement, and setting of quality standards. A man does not become a trio member unless he has shown the skill and tact to do this part of the job.

Trio Reports.—Each two weeks, each development trio presents a progress report—either oral or written—to management. If the trio runs into trouble, the executives take immediate action.

Each Monday, Cotton (1970) reports, selected members of the research department meet with the president, the head of one of the divisions, and representatives from advertising, market research, production, purchasing, and engineering to review progress and expedite product development.

In this way research people are made to feel part of the team and responsible for their share of the development effort. The attention by top management tends to keep projects in line with corporate objectives, bolstering teamwork and morale.

Each research supervisor meets every three months with laboratory management—the director of research, the associate director of research, and the manager of research administration. During these meetings, results are compared with plans. New plans are made or finalized in response to company needs.

In line with activities indicated in Chap. 6, the national company has a Research Committee composed of the chairman of the board, the president, and the vice-president of research. Programs and progress are reviewed monthly. Several times a year this committee examines in depth one area of the business and the research program in that area. When the meeting is concerned with a specific division, the head of that division joins in the meeting. Basic policies and directions are established during these meetings.

Management of product development takes management—time, support, review, funding, and guidance. As the above indicates, management, even in a global corporation, gives the time that research laboratory activities demand to achieve results.

Big Systems R&D

How can a multibillion dollar global food corporation with a global centralized corporate laboratory complex, manage hundreds or even thousands of projects for dozens of national companies? Research management techniques have also been evolving. It is now possible to keep abreast of these projects being undertaken by laboratory units of various sizes, containing hundreds or thousands of research personnel. They do this by drawing upon operations research techniques.

Operations research has been noted earlier to be defined as an approach to the solution of complex problems related to the direction and management of large systems of men, resources, and functions.

It is characterized by a model of the system under study.

It includes such factors as chance and risk, by means of which the outcomes of alternative decisions, strategies, and controls can be predicted.

It provides the basis upon which decisions can be made by management. The first step is to make a plan, the object of which is to give a clear indication of how the overall aims may be achieved.

PROJECT MANAGEMENT BY CPM AND PERT

Moder and Phillips (1964) distinguished research project management from production management primarily by the nonrepetitive nature of the work defined as a "project." A project, as defined here, is usually, but not always, a one-time effort. Although similar work may have been done previously, it is not being repeated on a production basis. Consequently, the project manager must plan and schedule largely on the basis of his experience with similar projects, applying judgment to the particular conditions of the project at hand. During the course of the project, he must continually replan and reschedule because of unexpected progress, delays, or technical conditions.

According to Moder and Phillips, until just a few years ago there was no generally accepted formal procedure to aid in the management of projects. The development of critical path methods (CPM) in 1958-1959 has provided the basis for a more formal and general approach toward project management. They found that CPM involves a graphical portrayal of the interrelationships among the elements of a project, and an arithmetic procedure which identifies the relative importance of each element in the overall schedule. Since then CPM has been applied with notable success to large-scale research and development programs. Applications have also included production planning, where the introduction of a new product or changeover to a new model is a project. (See Table 8.2.)

TABLE 8.2

SOME GOVERNMENT AGENCIES THAT HAVE INCLUDED THE USE OF CPM OR PERT IN CONTRACT SPECIFICATIONS

U.S. Government Agency	Method Required
Department of the Navy	
Bureau of Naval Weapons	PERT
Special Projects Office	PERT, cost control
Bureau of Yards and Docks	CPM, cost control
Bureau of Ships	CPM and PERT
Department of the Air Force	
Air Force Systems Command	PERT, cost control
Ballistic Systems Division	PERT, cost control
Electronic Systems Division	PERT
Department of the Army	
Corps of Engineers	CPM, cost control
National Aeronautics and Space Administration	CPM and PERT, cost control
Federal Aviation Agency	CPM
General Services Administration	CPM

Source: Moder and Phillips (1964).

In all projects one is concerned, to a greater or lesser degree, with developing a workable plan. Also, one is interested in scheduling an acceptable time span, and finally controlling the conduct of the scheduled work. One should first plan the work and then work the plan.

With respect to planning and scheduling, the aim is to conduct the program so that the cost and time required to complete the project are properly balanced, and excessive demands of key resources are avoided. The control function is concerned with monitoring the expenditure of time and money in carrying out the scheduled program, as well as the resulting product quality or performance.

Development of the Network Plan Concept

The heart of CPM, Moder and Phillips (1964) found, was a graphical portrayal of the program's activities leading to the end objective. This graph is called a network. The network concept has been used for more than a decade.

In 1958, project networking was defined concurrently by 2 research teams, 1 developing Program Evaluation and Review Technique (PERT) and the other the Critical Path Method (CPM). The features generally associated with CPM and PERT are applied as appropriate to a project, regardless of whether "CPM," "PERT," "Network Analysis," or another term is used to describe the general approach.

Moder and Phillips reported that the objective of the PERT research team was to design a management control system for the development of the Polaris missile system, a program whose activities were subject to a considerable degree of uncertainty.

They found that the objective of the CPM research team was to determine how best to reduce the time required to perform routine plant construction work, and hence to minimize the sum of the direct and indirect costs. Hence, the system developed by this research team had as one of its main features the planning and scheduling of a project with emphasis on performance time and total costs.

The basis of both CPM and PERT is the project network diagram, which is primarily designed to control the time element of a program, as depicted in Fig. 8.4. Here, the bar chart portrays the major activities, their scheduled start and finish times, and their current status. The important ingredients added by the network are that the dependencies of activities are noted, and detailed definition of activities is made. These two points will be discussed below.

In the network they attempt to demonstrate these points (Fig. 8.4); the lines denote activities which usually require time, manpower, and facilities to complete. Each activity originates and terminates in a unique event; time flows from the tail to the head of each arrow. Events denote a point in time; their occurrence signifies the completion of all activities terminating in the event in question. For example, the occurrence of event 7 signals the completion of activities 3-7 (3 → 7) and 6-7. Dashed line arrows, called dummies, show precedence rela-

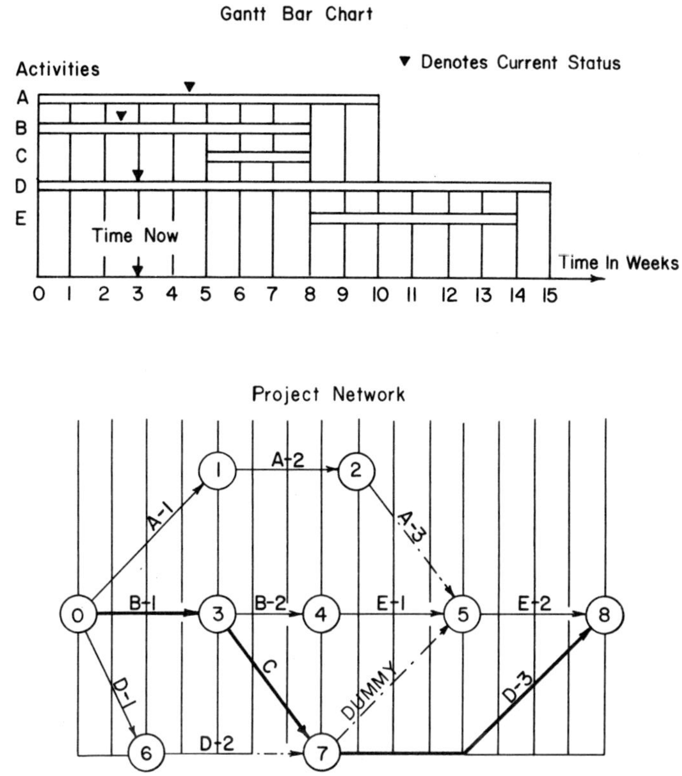

Courtesy of Moder and Phillips (1964)

FIG. 8.4. COMPARISON OF GANTT BAR CHART AND PROJECT NETWORK

tionships only; they require no time, manpower, etc., to perform. Such a relationship is shown by activity 7-5. This networking scheme is the most widely used today, according to Moder and Phillips. It is called activities-on-arrows.

In Fig. 8.4, activities A-3, D-2, and E-2 have the last portion of their arrows dashed. This denotes activities expected to be completed prior to the occurrence of their succeeding events. The paths along which these activities lie are referred to as "slack paths." They note, for example, the expected time to perform activities A-1, A-2, and A-3 is 4 + 4 + 1 = 9 weeks, as indicated by the time scale. The time interval between the occurrence of the initial and terminal events of this path, i.e., events 0 and 5, is 11 weeks. Thus, this path has 2 weeks of "slack." Activities B-1, C, and D-3 have no dashed portions; the sum of their expected performance times, 15, is the same as the time interval between the occurrence of events 0 and 8. For this reason, this path is referred to as the "critical path"; it is the longest path through the network.

The majority of critical path applications, they report, contain the following phases: the preparation of the network diagram, the estimation of expected time to perform each activity, the computation of the critical path schedule, and the interpretation of results.

Critical path procedures require a greater planning effort than is necessary with conventional methods. Miller (1962) states that these procedures may double the planning costs. But this cost is easily justified by concentrating attention on the critical path and avoiding unnecessary expenditures such as across-the-board overtime. This is of considerable importance, since he found the activities which lie on the critical path generally comprise only about 15% of the total project activities.

Further, the basic network provides a basis for detailed analysis of the overall objectives of a project and its resource restraints. Moder and Phillips (1964) found that the variations and supplements to the basic critical path concepts deal largely with analysis of the uncertainties in the schedule, the time-cost relationships, the efficient utilization of labor and equipment, and the control of expenditures. In several of these areas, the potential advantages of the critical path approach are only just beginning to be realized in practice.

It is useful to recognize that there are basically two different types of project activities and all shades in between. Activities which perhaps have never been performed before are called variable activities, and are characterized by a relatively large variance in their performance time. Examples of this type of activity are the development of new items. They show an example of a variable type of activity on the top in Fig. 8.5. A deterministic type of activity is accurately known; its variance in performance time is negligible, as shown in Fig. 8.5. Examples of deterministic activities are well-established, analytical costs.

They found that programs comprised primarily of variable activities employ the PERT version of critical path methods.

On the other hand, programs comprised primarily of deterministic activities (such as maintenance projects) utilize CPM.

ALLOCATION OF RESOURCES

The network plan displays the project activities and their order of performance. By the addition of time estimates, project activities can be calendar oriented. Calendar orientation provides a means of scheduling the utilization of personnel or resources. An example of such a check is shown in Fig. 8.6. If the peak demand for this resource occurring in the time interval, t_1 to t_2, is judged excessive, CPM provides a means of replanning and scheduling the project to eliminate this peak.

The time-cost trade-off problem is directed to the task of determining a sched-

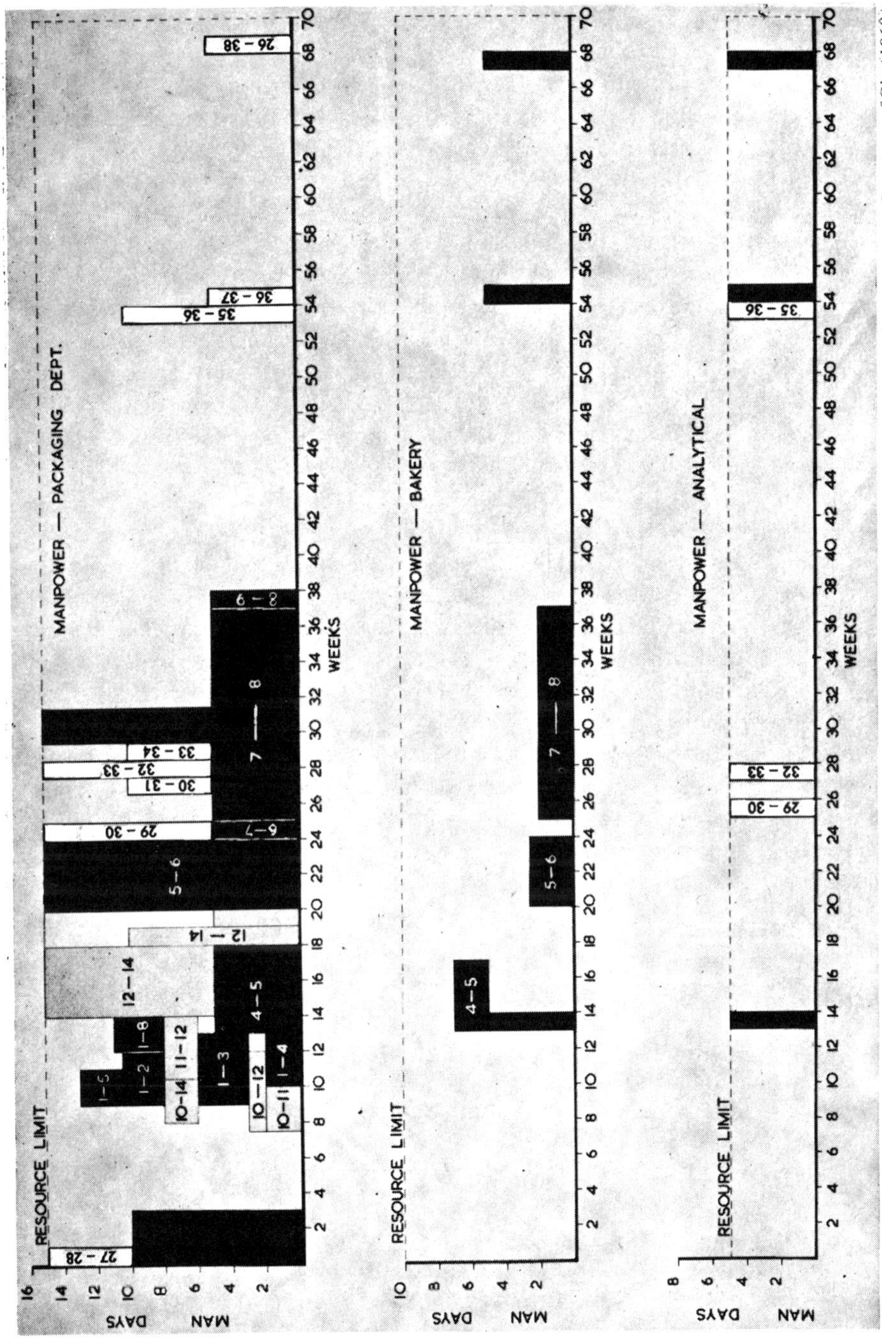

Courtesy of Sly (1968)

FIG. 8.8. RESOURCE LOADING CHART

Sly (1968) found that the process of preparing a network for a research project is a useful exercise for the project leader since it encourages him to make a thorough preliminary analysis of the work elements of a project.

The data contained in a network can be used to prepare an activity shedule, indicating by date the start and estimated finish of work to be done. The activity schedule has two useful functions: (1) It sets clearly the responsibility for each particular activity. (2) It enables anyone concerned to see at a glance whether the project is on schedule.

Subnetworks are also established as required. Eventually, 2 lists can be assembled for all projects, 1 for events and 1 for activities. The list of activities is in order of ascending latest starting dates, regardless of the network involved. The list of events for each network shows the earliest starting dates. The project to which the event belongs is identified. Such lists can be used to prepare a resource allocation chart (Fig. 8.8) that indicates the working demand on available facilities. The resource loading chart can clearly indicate the points in time at which more work can be taken on and when overloading will occur.

SUMMARY

Research requires management and control. All the techniques and schemes described to do these can accelerate the flow and analysis of information. The importance of this in being able to correct projects which are overrunning in cost and time at the earliest possible moment cannot be too highly stressed, particularly since scientific manpower and equipment are very expensive.

However, let us not confuse the ends with the means. Methods of approaching research laboratory management, however ingenious, can be justified only if they genuinely help us to attain objectives. An industrial research organization should provide a company with new and improved products and processes upon which its future profitability may be soundly guided. How it does this will be the subject of the next chapter.

BIBLIOGRAPHY

Allen, J. A. 1967. Scientific innovation and industrial prosperity. Elsevier Publishing Co., New York.

Anon. 1970. R&D management. Harvard Business Rev. Reprint Serv., Soldier's Field, Boston.

Bass, L. W. 1967. The planning of innovation. Chem. Ind. 1671-1675, Oct. 7.

Bass, L. W., and Old, B. S. 1967. Technical programming for industrial companies. Formulation of Research Policies. Am. Assoc. Advan. Sci., Washington, D.C.

Cotton, R. H. 1970. Management of product development in a large diversified corporation. Food Technol. *24,* No. 4, 36-38.

Judson, A. S. 1966. A Manager's Guide to Making Decisions. John Wiley & Sons, New York and London.

Hilton, P. 1967. New Product Development. Prentice Hall, Englewood Cliffs, N.J.

Miller, R. 1962. How to plan and control with PERT. Harvard Business Rev. *40,* No. 2, 93-104.

Moder, J. J., and Phillips, C. R. 1964. Project Management with CPM and PERT. Reinhold Publishing Co., New York.

Shepard, H. A. 1955. Research teams in industrial laboratories. MIT School Ind. Management, Cambridge, Mass.

Sly, W. H. 1968. What is involved in the management of R&D? A British Evaluation. Food Technol. *23,* No. 1, 41-46.

Wilson, E. B. 1952. An Introduction to Scientific Research. McGraw Hill Book Co., New York.

SECTION II

Managing People and Innovation

CHAPTER 9

Nature of the Creative Function in Industry

Having established the real objective of new product development in a company, and explored the marketing related functions, let us now turn to the research laboratory aspects of the function and the nature of the people engaged in it, with the goal in mind of maximizing the returns on investments. Industrial laboratory efforts, too, must be profit-oriented.

Common sense tells us that sciences are fact-oriented and deal with physical matters and material things in their relationships one to another. The function and the man, in the practical commercial world in which we work, take on an added dimension. If we are to be successful, the function and the man require effective creative flows.

The function demands and the man therefore must possess venturesome imagination in the application of his scientific and technical knowledge. He is required to be as creative as an imaginative composer, writer, or artist.

Experience reveals that much of the work demanded in new product development by large modern-day corporations often results in an attempt to force creativity. True creativity can rarely be forced to produce a real innovation; it is easily stifled. Therefore, corporate managements must provide as uninhibited an atmosphere as possible in which stimulating free thought and imagination can occur.

At the same time, management must set forth broad guidelines, within which it expects creative product development. Obviously, it does little good to simply ask people to create—create what, for what purpose, and to what end? Therefore, companies tend to think in terms of channeled creativity, which, when uninhibited in its field, will generate ideas, concepts, and products in the most productive and abundant manner.

In this chapter we will explore certain aspects of the nature of the function of creative people and the influence of large organizations on them.

THE NATURE OF TASKS

Bass (1967) found, as a result of extensive analysis, that the natural sequence of steps in a technical undertaking is the following: first, there is a groping for an idea within one's field of interest; its merits are next examined by some type of exploration, initially by mental processes, often followed by preliminary experimentation; if the conclusions are favorable, a tentative goal is set; an analysis, with varying degrees of imagination and discrimination, is made that can be used to attain the objective; criteria are adopted for evaluating success; work is undertaken; results accumulated; analysis further reveals goal achieved and the project is brought to successful conclusion.

In fundamental research with a high degree of freedom, the goal may be redefined many times to reflect the changes of interest of the investigator. The sequence of informal planning of the attack becomes much foreshortened in the case of more routine, short-term types of work. It expands into increasingly flexible paths as studies approach basic science.

Technical activities in support of new product development can be broken down into arbitrary classifications which are often used; namely, quality control, technical service, product and process improvement, product and process development, innovative product and process investigations, long-range industrial research, and fundamental research. These categories are not, of course, sharply divided but form a continuum and merge gradually into one another. The points are summarized in Table 9.1

TABLE 9.1

SUMMARY OF INNOVATIVE CHARACTERISTICS OF TECHNICAL ACTIVITIES

Type of Activity	Goals	Character of Methodology	General Criteria	Opportunities for Planning Innovation
Quality control	Rejection of sub-standard products	Standardized tests	Adherence to manufacturing and marketing specifications	Extension to managerial control and to demonstration of needed improvements in raw materials, products, or processes
Technical service	Solution of immediate practical problems	Chiefly technical "know-how" and expertise	Direct application to operating procedures	Proposal of short-term changes for systematic betterment of manufacturing procedures and improvement in utility of products to customers
Product and process improvement	Up-grading of existing technology	Practical methods for introducing more effective technology	Demonstrable utility for improving manufacturing processes and for satisfying customer needs	Systematic improvement of company's competitive position in efficiency of manufacturing and acceptability of products
Product and process development	Extension of technology and marketing in present areas	Technology closely related to present operations	Compatibility of new technology with existing manufacturing and marketing skills	Critical evaluation of applications of present skills to extension of market potential
Innovative products and processes	Creation of radically different products and processes which will give competitive advantages	Technology within range of application under assumed future conditions	Concepts conforming to corporate long-range plans and resources for expansion of the business	Advanced technological concepts which when implemented will provide a basis for industry leadership
Basic industrial research	Establishment of channels for coupling new scientific discoveries with long-range industrial interests	No limiting restrictions except that technology could conceivably be developed to apply the information	Stimulation of research and development into application of basic changes in technology	Forecast of breakthroughs in technology which will lead several years hence to innovative products or processes
Fundamental research	To create new knowledge	No restrictions whatsoever	Satisfaction of the creative urge	Planning of programs which will lead to filling important gaps in knowledge

Source: Bass (1967).

In industry, many large firms support the entire range of categories, but the relative emphasis varies widely in different organizations.

RESEARCH PEOPLE AND CREATIVITY

The personal characteristics of people to be discussed below are limited to those related to our subject, new product development, and for which some empirical evidence exists of relationships to innovative research outputs.

To begin with, it is clear that there are personal characteristics which reveal people to be orientated more toward applied than basic research. A summary is shown in Table 9.2. As seen, some of the best documented differentiating characteristics are clustered in the terms "localite" and "cosmopolite." The sociological-psychological factors related to research innovativeness are roughly comparable to those found in studies of other innovativeness.

TABLE 9.2

VARIATIONS IN PERSONALITY OF THE RESEARCHER ASSOCIATED WITH APPLIED AND BASIC RESEARCH SKILLS

	Applied Researcher	Basic Researcher
1.	Is a "localite," that is, high attachment to employing institution	Is a "cosmopolite, that is, high attachment to own discipline
2.	Enjoys on-the-job social relations as ends in themselves	Has little concern for social relations
3.	Has high concern for the welfare of the organization's target population	Not particularly welfare oriented, regardless of personal background
4.	Has maximum interest in the research at point when findings are put to use	Has maximum enthusiasm and interest in research at point findings
5.	Prefers fixed hours and little talk of science on the job	Likes to talk science at all times
6.	Prefers jobs with socially compatible colleagues and technically competent supervisors	Emphasizes technical competence of colleagues
7.	Emphasizes salary and other economic rewards; publication opportunities less important	Emphasizes publication opportunities; salary less important
8.	Likes collaborative research	Tends to prefer working alone
9.	Doesn't believe in "narrow specialization"	Thinks specialization necessary
10.	Enjoys working on problems assigned to him if they have clear utility	Enjoys working mainly on problems of self determined intellectual interest

Source: Liston (1970).

In particular, the innovators tend to rank high in socioeconomic status, and as either not well integrated into a tight local social system or well integrated into a loose system in which innovation is itself a norm (Menzel 1960).

Studies by Straus (1961) revealed that high research performance is associated with the researcher's having primary loyalty to his discipline rather than to the organization in which he is employed. This is true when the institution is a

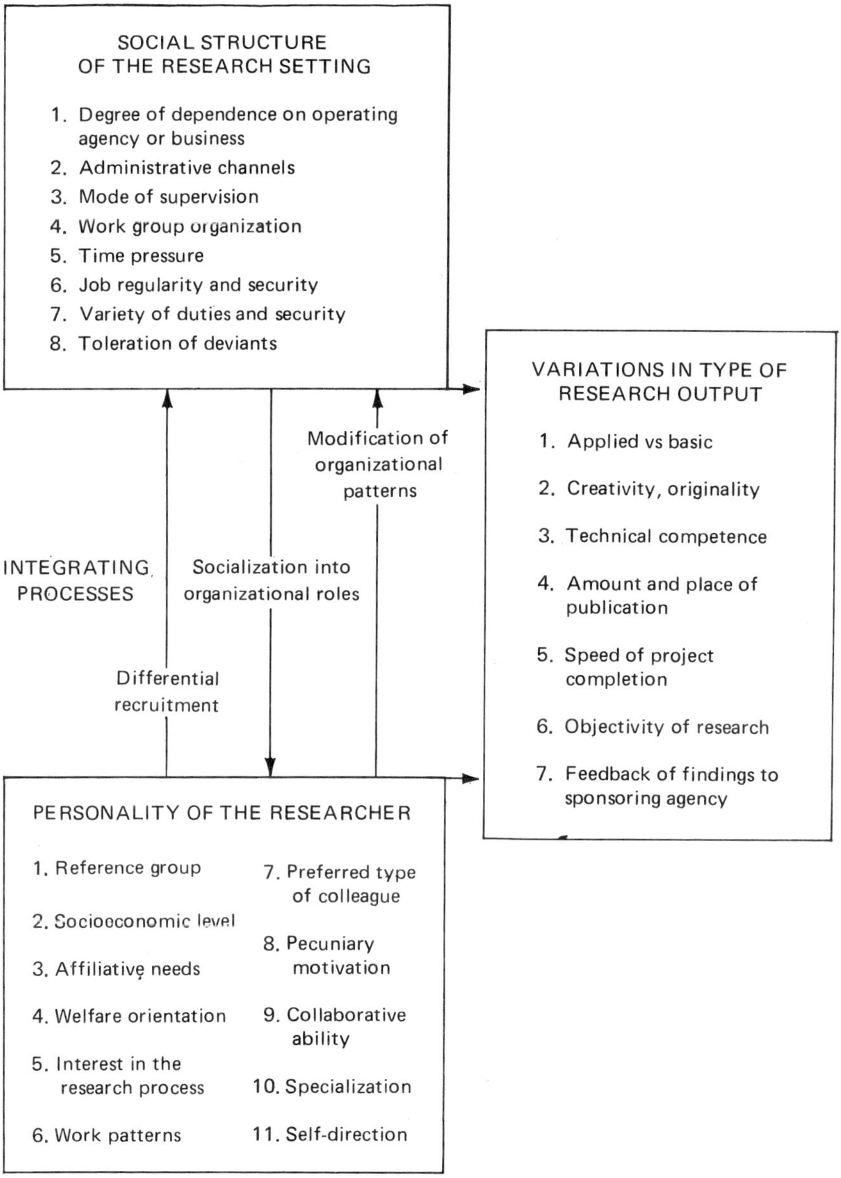

Courtesy of Straus (1961)

FIG. 9.1. SOME STRUCTURAL AND PSYCHOLOGICAL DETERMINANTS OF VARIATIONS IN TYPE OF RESEARCH OUTPUT research organization. It is markedly so when the institution is an operating agency or business. There are many reasons for this: the institutional loyalty of the localite makes him more willing to spend time on matters which are important to the institution (for example, committees) but of no scientific value; the localite's more complete acceptance of the organization's point of view makes it difficult for him to espouse novel approaches; and, a cluster of factors including the amount of communication and contact outside the institution, willingness to change jobs if the research opportunities turn out to be unfavorable, acceptance of the organization's time schedule even if it means lower quality work, and degree of willingness to take research risks which might result in locally known failures (Storer 1961).

Affiliative needs is another of the factors underlying the local versus cosmopolite distinction of research people. In particular, the low affiliative needs of the basic research cosmopolite permits him to be detached from the employing institution. Davis (1956) comments that these men are apparently able to obtain their affiliative needs for an entire year from a single annual scientific meeting, with perhaps an exchange of a scientific reprint or two as refreshers.

To summarize these contrasting characteristics of the personalities of applied and basic researchers, studies reveal that basic research-oriented investigators tend to get more rewards out of research itself. The applied researcher tends to derive his satisfaction out of things related to research, such as satisfying social relationships and the knowledge that someone will be able to use his findings directly. By bringing such studies to bear on the issue, some criteria emerge of importance in evaluating the organization for and planning of research in the various areas of new product development and its support. The types of research personnel needed and the criteria by which progress is evaluated, should be consistent with the central functions and objectives of the organization.

STRUCTURAL AND PSYCHOLOGICAL DETERMINANTS

Most of the studies on "research on research" reveal a marked contrast between the patterns of research organization which seem to work best where immediate utility is the critical requirement, and the organizational patterns which seem to work best where creativity, irrespective of immediate utility, is desired. Consequently, for each of the structural factors listed in Fig. 9.1, Liston (1970) tried to indicate in Table 9.3 variations which would maximize immediate utility, and others which maximize creativity.

The accompanying diagram from Straus (1961) (Fig. 9.1) represents three major dimensions of the research situation to be considered if particular types of output are to be achieved effectively. The first of these is the social structure of the research setting as indicated in the box at the upper left of the diagram. The second is the particular type of research output desired as characterized by such objectives as those listed in the box at the right. Third, both the nature of the

TABLE 9.3

RESEARCH ORGANIZATION STRUCTURES WHICH MAXIMIZE IMMEDIATE UTILITY OF OUTPUT VERSUS PATTERNS WHICH MAXIMIZE CREATIVITY

Immediate Utility	Creativity
1. Within-agency locus	University or grant research
2. Vertical or hierarchical structure	Horizontal or pie structure
3. Normal administrative supervision	Laissez-faire supervision
4. Well-established work teams	Individual research or ad hoc research teams
5. Emphasis on project completion within specific time	Tolerance for delays and willingness to change the direction of research
6. Security, regularity, limited working hours	Competitiveness, self-determined work hours and work load
7. Full-time research on a single project	Variety of duties, especially those involving teaching and contact with scientists working on diverse problems
8. Emphasis on teamwork and cooperativeness	Toleration of the social and scientific "oddball"

social structure and the type of research output make certain personality characteristics of the researcher highly desirable, others less so or even undesirable. To get the study planned, carried out, reported, and evaluated, three integrating processes tend to take place, namely: differential recruitment, socialization of personnel into organizational roles, and modification of organizational patterns.

Each of the items listed in the upper left box in Fig. 9.1 was chosen by Straus because there was empirical evidence indicating some aspects of its effect on research output. Undoubtedly, there are other research organizational factors which also influence output.

ELEMENTS OF SOCIAL STRUCTURE

Table 9.3 corresponds to the upper box of the figure and was set up by Liston to point out some of the differences between social structures which are oriented primarily toward immediate utility research and those aimed mainly toward basic creativity.

Starting with Item 2 of this table, we find a factor of wide applicability—a vertical administrative channel versus a horizontal or pie structure. Meltzer and Salter (1962) found that the productivity of the research people they studied decreased as the number of supervisory levels in the research organization increased. Moreover, this tends to occur in both large and small organizations. Similar conclusions were inferred from Davis' study (1956) of scientists in federal agencies. No empirical studies seem available for applied research organizations. However, both Storer (1961) and Polanyi (1956) argue that such an effect does not occur in applied research. Why not? An explanation is pertinent here.

Polanyi (1956) suggested that theoretical research has certain self-regulating mechanisms comparable to the price mechanisms in a competitive market. Applied research lacks these and therefore needs a unified organizational structure to give it direction. More specifically, basic research is somewhat like the economy in that it has a public medium of exchange. Just as the entrepreneur contributes goods and is rewarded by money, so the basic researcher creates knowledge for the scientific system and is rewarded by esteem. To reap this reward, his new knowledge must fit into the system of the science, as indicated by publication in a scientific journal.

However, Liston (1970) thought these self-regulating mechanisms usually do not work for research on applied and new product problems. First, much applied research is not published except perhaps as patents. Otherwise, the new information is for internal use by a a captive audience which generally lacks the specialized knowledge to evaluate it. Second, applied research has no built-in system to guide the selection of research problems comparable to the role of theory in basic science. This suggests that it is just as legitimate for the business organizational hierarchy to select the applied scientist's research problem as it is for the scientist himself to select it. But in basic research, it seems reasonable that only the scientist knows the theory well enough; hence, he, quite rightly tends to reject the right of an administrative hierarchy to determine the course of his research.

Most of the other organizational contrasts shown in Table 9.3 are self-explanatory and only brief comment need be made about them.

Specifically in Item 4, work group organization, Liston (1970) found support in a study by Shepard (1955) that the longer a research group worked together, the lower its performance rating. This is the opposite of what might normally be expected, since work group efficiency should increase as the members learn to coordinate activities more precisely. Why this result? The best guess by those involved seemed to be that it is a function of the stimulation provided by the divergent approaches of a newly formed group. Each member of the team in the beginning tends to become better trained by learning from the others. Secondly, the superior performance of the newly-formed groups may stem from the solutions which emerge from attempts to reconcile and integrate the divergent ideas and techniques of the new team members.

FULL-TIME/PART-TIME PROJECT EFFORTS

In relation to full-time development versus a variety of duties and contacts (Item 7, Table 9.3), Meltzer found that full-time output goes up steadily until the point at which about 60% of one's time is spent on research, and then output declines. Moreover, although total productivity increases up to the 60% point, productivity per unit of time decreases from the start. In short, 2 half-time researchers do more than 1 full-time person. Liston felt the explanation was in the notion that new ideas, new outlooks, and new ways of doing things—all providing a richer source of material to draw upon in the creative process and the necessity of reconciling these divergent viewpoints.

In practice, it has been resolved by generally assigning more than one research project per scientist or technologist, and assigning lower priority to the other project(s).

BASIC VERSUS APPLIED

Returning to Item 1, in Table 9.2 relating to degree of dependence on an operating business, it becomes clear from the other items in the Table that a within-business research setting, organized along lines of administrative rationality, is conducive to a relatively rapid and assured output with a greater likelihood of having application relevance. However, this may be gained at the cost of innovativeness or creativity.

On the other hand, the loosely organized, individualistic, and competitive emphasis of academic departmental systems seems to be particularly well suited to maximizing creativity and productivity, but a productivity which will not necessarily have any direct application value. It is a paradox worthy of attention in new product development.

Socialization of researchers into the values of the organization and modification of organizational practices go on continually. These provide the integrating processes which allow crossing of lines between contrasting needs of research organizations and research persons. It is also worth noting that organization structural pressures and constraints are very strong and tend to convert the person with a basic research orientation to an applied one. This is particularly true today as the demand for sound applied research appears to be increasing. As demand presses heavily on resources for research, this question arises: How much should basic and theoretical research be sacrificed in order to give higher priority to applied studies, and vice versa?

Unfortunately, no one answer can be given the question which would have meaning for corporations as a whole, particularly in this day of diversified corporations and the nature of individual companies. But, insight can be gained from another direction by examining what the successful corporations do.

One way to begin is to look at the general investment levels in R&D of various kinds of industries. Parsons (1963) found that these range from upwards of 35% of gross sales for drug and electronic firms, to 5-10% for various chemical and petroleum-based companies, to 0.2-1.5% for basically food companies.

He found that the average of the investments in R&D in the major food corporations in the United States amounted to about 1.0% of gross sales. For a billion dollar company, this would yield about 10 million dollars for the R&D function.

Distributed between basic and applied research, or between fundamental and product-oriented research, in the food industry it is generally believed that about 20% is being devoted to fundamental and about 80% to product-oriented activities.

To bring about the R&D functions in a major, diversified corporation operating in the mass market, consumer-based food industry, with between 500 million and 1.5 billion dollars of gross sales, it takes between 150 and 800 R&D personnel respectively.

ORGANIZATIONS AND QUALITY OF LIFE

All institutions have power, all exercise power, and all need to take responsibility for their actions. Whoever assumes responsibility asserts authority. Conversely, one is obviously responsible for whatever one has authority over.

To take responsibility where one has no authority is usurpation of power.

What is proper exercise of authority? What impacts do organizations have because of their functions? Drucker (1968) suggested the following: (1) All organizations have an impact on society in order to carry out their missions. (2) All organizations have to be somewhere. They are localized. They have local impact, wherever they operate. (3) Organizations employ people. They thus have an impact on people. They exercise authority over them. (4) Impacts on people and communities are not the purpose of organization. They are incidental to it.

It is thus clear that we would not have organizations if we knew how to obtain performance without them. This being the case, it would be useful to have some rules of the road to guide people in their relations with organizations and vice versa. Those gleaned by Drucker are:

(1) Organizations should limit their impacts on people as much as possible.

(2) Organizations should limit their impacts on all other social-community interfaces.

(3) To demand loyalty from employees under all conditions is wrong, he reports. It is one thing to receive affection, gratitude, friendship, mutual respect, and confidence from employees. But, it must be earned. Otherwise, these demands are incidental to the tasks of organizations.

(4) One duty of organization is to anticipate its impacts. What impacts ahead are likely to become social problems? They need careful thought.

(5) Another duty of organization is to try to prevent undesirable side effects of its impacts.

(6) Whenever an undesirable impact is not prevented, it will boomerang. It leads to regulation, punitive laws, and outside interferences.

(7) Whenever business leaders anticipate an impact on business, and have thought through its prevention or treatment, their proposals have been accepted. There are many examples of this if we think about the matter.

(8) Whenever the institution shrinks back, pretending "someone will not permit us," it pays a heavy price in the end. While people will forgive blindness, they will not forgive failure of an organization to act in its own best knowledge. This is organizational cowardice, he suggests.

(9) Voluntary efforts in which everyone has to do something that in the short-run is very risky and unpopular has never succeeded.

(10) Ideally, an organization converts into opportunities for its own performance the satisfaction of social needs and wants, including those created by its own impacts.

(11) Every organization is expected to be an entrepeneur.

(12) It is ethical for business to convert into business profit the satisfaction of social needs and wants, Drucker states.

(13) To seek opportunity is the ethics of business. Social awareness is organizational self-interest.

(14) The needs of society, left unfilled, lead to social diseases, he believes, which reduce the quality of life.

(15) Leadership in the society of organizations is not a function of position in society. It is a function of the position in society's organizations, composed of diverse pluralist organizations.

(16) Society expects executives of major businesses to anticipate social problems and to work at their solutions.

(17) The best way for business, Drucker finds, is to turn social needs into profitable new business.

(18) Society holds leaders of organizations responsible for the quality of goods and for the quality of life, he finds.

These are some of the fundamental principles of organization and its management, and they are worthy of serious consideration.

In line with the creative role in new product development, it appears that we must strive to develop a new generation of creative contributors. Each should be encouraged to ask: (1) What do I aspire to contribute? (2) How can I make this organization serve my ends and my needs? (3) How can I make this organization enable me to perform, to achieve, and to contribute? (4) How can I use this organization intelligently, purposefully, deliberately, and responsibly?

Such people will make organizational contributions that bear fruit.

Why have we dwelled to such length on the above? Because the accumulating literature on the subject is replete with case histories where the common sense of matters was ignored. Recall the examples and statistics discussed earlier in Chap. 5, and the findings of Angelus and others.

If we are to build on the past to improve the success ratios in new product development in the future, we must find answers; and these have been found. It remains to apply them effectively.

Further, the world must be seen an unfolding in new opportunities. As noted at the outset, one discovery always leads to another and this yields still another in an unending wave of new innovations. A good deal depends on our attitudes in such matters as to whether or not we focus on the past or build firm foundations for the future. Having 20-20 vision in hindsight may be useful for an historian, but for people involved in new product development, it is foresight that matters.

People involved in new product development must also be committed to the task. Not everyone is suited for the task. Those that are are clearly visible within organizations.

Now, let us turn to the future, building on the past.

BIBLIOGRAPHY

Angelus, T. L. 1970. Improving the success ratio in new product development. Food Technol. *24,* No. 4, 29-36.

Anon. 1970. R&D Management. Harvard Business Rev. Reprint Serv., Soldier's Field, Boston.

Bass, L. W. 1967. The planning of innovation. Chem. Ind. 1671-1675, Oct. 7.

Baumgartel, H. 1956. Leadership, motivations and attitudes in research laboratories. J. Social Issues *12,* 24-31.

Davis, R. C. 1956. Commitment of professional values as related to the role performance of research scientists. Ph.D. Dissertation, Univ. Michigan, Ann Arbor.

Drucker, P. F. 1968. The Age of Discontinuity. Harper & Row, New York.

Fredrickson, E. B., and Lawson, E. W. 1970. The role of technology transfer in product development and marketing. Res. Management *13,* No. 4, 265-272.

Liston, M. I. 1970. Some basic requisites for research productivity. J. Home Econ. *62,* No. 4, 234-245.

Meltzer, L. 1956. Scientific productivity in organizational settings. Social Issues *12,* 32-40.

Meltzer, L., and Salter, J. 1962. Organizational structure and the performance and job satisfaction of scientists. Am. Social Rev. *27,* 351-362

Menzel, H. 1960. Innovations–a survey. Am. Social Rev. *25,* 704-713.

Parsons, L. B. 1963. Personal communication. Ridgewood, N.J.

Polanyi, M. 1956. Pure and applied science and their appropriate forms or organization. Dialectics *10,* 231-242.

Shepard, H. A. 1955. Research teams in industrial research laboratories. MIT School Ind. Management, Cambridge, Mass.

Storer, N. W. 1961. Science and scientists in an agricultural research organization. Ph.D. Dissertation, Cornell Univ.

Straus, M. A. 1961. Social-psychological aspects of extension research organizations. Natl. Extension Res. Seminar, Purdue Univ., Lafayette, Ind.

CHAPTER 10

Financial Business Decision Guides

If new products were not profitable, product development would not be the vital function it is in corporate life today. Therefore, an important element in new product development is, obviously, profitability consideration. Despite awareness that profitability is the vital factor in the decision on a product, it is often one of the least understood. Incremental profitability is the basis for the vitality of a new product, and it affects the overall growth and health of the corporation. The financial function in new product development is no less important than any other area of specialization. With this by way of introduction, let us explore some of the important business concepts, starting with finance as related to new product development, then move on to business guides.

FINANCIAL ASPECTS

According to Cavalier (1967), the financial function is clearly to make new product goals mesh with one of the company's major goals—the maximization of profits, and, in turn, shareholders' returns on investments. Included in this definition of new products is acquisition or merger to accomplish goals.

Volume Is Not Profitability

He found that some people tend to ignore the total objective of a corporation by equating sales volume with profitability. Sales volume as a measure of success is an oversimplification that can be misleading and harmful. This can be seen in Table 10.1.

TABLE 10.1

NEW PRODUCT FINANCIAL STATEMENT

(Good New Product) X (Efficient Manufacturing) X (Effective Marketing) = (Profits)

This statement can be translated by rearrangement and substitution as follows:

$$\text{Good New Product} \times \frac{\text{Sales}}{\text{Investment}} \times \frac{\text{Net Income}}{\text{Sales}} = \frac{\text{Net Income}}{\text{Investment}}$$

Source: Cavalier (1967).

One of the obstacles, he believes, to a better understanding of the financial function arises because of misconceptions that associate accounting with finance. While accounting seems a logical point of departure for understanding finance, the accountant is primarily interested in classifying data and is not necessarily concerned with the flow of funds. Accounting, he explains, focuses on the proper collection and presentation of the data. Once the data are collected and validated, financial people then can analyze and manipulate data in a mathematical sense to gain insights into the corporate health.

For our purpose here, assume that the accounting data used are accurate and properly classified according to accepted accounting standards. Next, one could group the data into a few classifications, as shown in the accompanying table (Table 10.2).

TABLE 10.2

STEP 1: TYPICAL (SIMPLIFIED) ACCOUNTING STATEMENT

Assets		Liabilities and Stockholders' Equity	
Current assets	$600	Current liabilities	$300
Plant and equipment	$300	Long-term debt	$100
		Total liabilities	$400
Other assets	$100	Stockholders' equity	$600
Total assets	$1000	Total liabilities and stockholders' equity	$1000

Source: Cavalier (1967).

Using such data, Cavalier (1967) developed a financial model that shows the flow of funds through a corporation. In devising the model, he used an input-output type of relationship. For example, we might need to know the source of funds in the company, the application of these funds, and some measure of the efficiency with which these funds are being used. Knowing these factors, one could compare the appeal of a new product against existing relationships of products within the company, as well as with similar factors for alternative products. This input-output rearrangement he shows as Step 2 of Table 10.3. In it, working capital is the difference between current assets ($600) and current liabilities ($300); the example would hold if we used thousands or millions. These funds can be used on a daily or short-term basis, as opposed to long-term funds fixed in plant and equipment. By combining current assets and current liabilities, one retains the equality between the source and the application of funds.

Cavalier defines capitalization as a combination of long-term debt plus stockholders' equity, and represents the source of long-term funds to the corporation. (Another term is invested capital). Step 2 in his model is static, since balance sheet items reflect valuations as of a certain day. But a corporation is an on-going entity, and this view is achieved by introducing the corporation's income statement into the model as he shows in Step 3, Table 10.4

TABLE 10.3

STEP 2: REARRANGEMENT TO ESTABLISH INPUT-OUTPUT FLOW

Source of Funds	($700)	Application of Funds ($700)	
Long-term debt	$100	Working capital	$300
		Plant and equipment	$300
Stockholders' equity	$600	Other assets	$100
Total capitalization	$700	Uses of funds	$700

Source: Cavalier (1967).

TABLE 10.4

STEP 3: ADDITION OF FIRM'S INCOME TO FINANCIAL MODEL

Total Assets =	Total Liabilities +	Stockholders' Equity
($1000)	($400) +	($600)

Capitalization			Income Statement
Long-term debt Stockholders' equity	=	Working capital Plant and equipment Other assets	Sales (Less expenses) Net income

Source: Cavalier (1967).

Unlike the balance sheet, the income statement represents the link between balance sheet periods.

Reviewing Cavalier's model, in Step 3, funds enter the corporation, are invested, and generate a given volume of sales. Since expenses are incurred in producing sales, these costs must be deducted to arrive at net income after taxes. The problem then, he finds, is to balance the needs of the company for funds with the demand of the stockholders for dividends. Net income can be distributed in whole or in part to the stockholders; it can be retained in whole or in part by the corporation. He notes that, if the funds are retained they must be added to the stockholders' equity, since this money still belongs to the stockholders. This is shown as Step 4, Table 10.5.

Step 4 in his model shows how new funds flow through a company to permit the cycle to repeat itself, producing further capital. In order to translate new product marketing concept into the model, Cavalier suggests the following guideline: A good new product, efficiently manufactured and marketed, will generate profits. This can be stated as an equation, shown in Table 10.1.

TABLE 10.5

STEP 4: DISTRIBUTION OF NET INCOME

Capitalization			Income Statement
Long-term debt	=	Working capital Plant and equipment	Sales (Less expenses)
Stockholders' equity			
→Retained earnings		Other assets	
			Net income
←		Less Dividends	←

Source: Cavalier (1967).

The ratio, net income : investment is called the return on invested capital (capitalization).

Thus, using Cavalier's model approach, we can see the interrelationships of new product marketing and finance. In addition, this model stresses an earlier point: sales volume alone is not the sole criterion for determining the profitability of a new product.

PROFITABILITY OF NEW PRODUCTS

If we accept the data available, about 80% of all R&D efforts in new products are doomed to failure. Industry in the United States is now spending about $19 billion on R&D. If 80% of these funds are used for product research and 20% for basic research, as suggested in Chap. 9, the overall system is indeed wasteful. Much of the waste is attributable to new products that fail. One of the many reasons that dooms new products to failure is laboratory failure.

One of the ways in which laboratories can increase their efficiency and also gain a greater appreciation of the impact of their work is to utilize a decision guide. Goodman (1970) proposed a guide encompassing many of the critical facets. These guides are shown in Fig. 10.1.

Urgency and Risk

Priorities are established in the section of Goodman's decision guide dealing with urgency and risk. As an example, about 5 yr ago company A generated a new product idea for cat foods, packed in different tins, with a number for each day of the week placed on a series of cans. The animal was to receive a different variety of food each day of the week. The product was test marketed. More than 1 yr later, they decided to market the product nationally. By the time they got around to the decision, they found that company B had already gone national with a very similar type product. All that company B needed to do was to watch and follow the progress of company A's test. Then company B made a decision to enter

MARKETING DIVISION:	TITLE:	
LOCATION:	PROJECT NUMBER:	DATE:

I. PRODUCT DESCRIPTION

A. ____ New Product ____ Quality Improvement
____ New Size ____ Exploratory Research
____ New Flavor ____ Concept Testing

B. General product description, major uses, consumer advantages and competitive advantages where applicable: ____

C. Rationale, or justification for project proposal: ____

D. Specific Objectives: ____

II. URGENCY AND RISK

A. Urgency ..... ____ Normal development cycle
____ Longer than normal development cycle
____ Crash program - speed essential

B. Degree of Risk ..... ____ Virtually none
____ Moderate
____ Substantial

III. COMPETITION

When possible, for each known competitive or related product include brand or description, unit size and/or number of servings, package type, (carton, glass, etc.), unit retail price, estimate yearly volume, number of years on market, distribution pattern description, estimated yearly advertising or promotional expenditures; and general comments on product advantages and disadvantages: ____

IV. TECHNICAL INFORMATION

A. Manufacture of Product
1. a) Our Manufacture ____ b) Indicate Manufacturing Locations: ____
c) Outside Processors ____

Courtesy of Goodman (1970)

FIG. 10.1. NEW PRODUCT DECISION GUIDE

V. DISTRIBUTION

A. ________ Existing Sales Force

________ Additional Salesmen to be hired

________ Other

B. Warehousing and Handling Facilities: ________________________________

C. Comments on special manufacturing, shipping, storage, handling problems, etc.: ________________________________

VI. PROFITABILITY

A. Forecast (Incremental Only)

	19	19	19	19	19	19	19	19	19	19
Net Sales										
Variable Profit										
Advertising										
Promotions										
Selling Expense										
Division Profit										
Incremental Exp.										
Net Profit										

B. Expected unusual promotion vehicles: ________________________________

C. Will the proposed product or process allow royalty payments?

______ Yes ______ No

VII LIFE CYCLE

A. Estimated time for basic development ________________________________

B. Estimated time for test marketing ________________________________ (cum.)

C. Estimated time for national distribution ________________________________ (cum.)

D. Estimated time for maturity* ________________________________ (cum.)

*Softening of price structure, marginal return of marketing investment, proliferation of competition, decline of consumer preference for brand.

2. Is new equipment required? ______Yes ______No

If yes, complete below:

Incremental Investment	19__	19__	19__	19__	19__	19__	Total
1) Capital Cost							
2) Related Expense							
3) Working Capital							
Total							
Depreciation							

3. Proposed Supplier of raw materials for production: ______

Comment only if problem is anticipated: ______

4. Special manufacturing problems: ______

B. Finishing Supplies

1. Type of Container:

______ Jar ______ Can ______ Envelope

______ Other: Specify ______

2. Attach sketch or photo of package design, if possible.

3. Comment on technical packaging requirements: ______

C. Product Stability

1. Shelf life in recommended packages: ______

2. Container compatability: ______

3. Other stability problems: ______

VIII. PROBABILITY OF SUCCESS

	Technical	Marketing
0 - 25%	______	______
25 - 50%	______	______
50 - 75%	______	______
75 - 100%	______	______

IX. Using the above projections for expenditures and income, complete the attached forms to compute the Return on Investment and Payback Period and enter the results on the line below.

______ % R.O.I. ______ Years Payback Period

the area, create the product, and place it on the market ahead of company A. Clearly, company A miscalculated the degree of urgency and risk as they conducted their project, Goodman suggests.

Competition

Competition—whether external or internal—must be considered carefully. As a matter of corporate policy, Goodman found many companies create products which compete with each other internally.

An example of such a decision had to be made by one company when it developed a frozen dessert topping. Prior to the development, they had had a successful powdered topping on the market for about 8 yr. Both products would essentially compete in the same market. Fortunately, their decision to go ahead with the frozen product resulted in a very satisfactory performance for both products (Angelus 1970).

However, Goodman (1970) reported that this element of competition must also be considered if for no other reason than that the payoff in the profit and loss statement of a company can only be expressed in incremental terms. "Incremental" means the profit over and above which is the result of an action. It is the added amount of profit which results, beyond that of the already existing products, when the new one appears on the market.

External competition is also to be faced. A new product entry by a company in an area, in a market dominated by other companies, will find high potential sources of competition. Consideration of the longevity of life for the new product, pricing levels, volume levels, types of packaging, and even the level of quality itself require expert attention, he cautions.

Selling

Sales managers are like marketing product managers and like managers of R&D. They all believe in the work they are doing; they must. How objective are they about the potentials for their products?

Goodman (1970) found that national sales managers are almost always convinced that their sales force can sell any product. Yet, if a salesman has four lines to sell and he is given a fifth, he faces a real risk in his ability to efficiently sell that fifth product. Not only must he spend more time selling the new product, but he also faces educating his customer to a new product concept or a new version of an existing product. Quite often, additional people will have to be added to the sales force; thus diluting the profit impact of the new product.

Profitability

Profit and profitability are not the same, as we have already seen. As

Goodman (1970) sees the situation, profit is a very tax-oriented concept and essentially means it is the amount of money left over after subtracting all possible expenses from the revenues. The major problem in utilizing the word profit is that it is a very static concept. Besides that, the controller of a firm can employ the generally accepted accounting principles, the guidelines of the profession, to change the magnitude and the direction of profit. For example, the controller can accelerate depreciation, or in effect, retard it. He can decide to capitalize certain expenses and consider them as assets of the company. Or, he can choose to expense these items and consider them immediately as rêductions of revenue on profit and loss statements.

Goodman maintains that these are all subjective judgments and these transactions can be handled in more than one way. The choice that one makes will affect profit. For these reasons and many others, he considers the word profit as a futile point at which to begin to make action decisions about the efficiency and validity of research and developmental work. The preferred term is profitability. The concept employs only the incremental, economic impact of any revenue or expense items affecting the product or project. In essence, it says that no costs should be allocated to a new product activity which does not result from the existence of the activity itself. Many costs cannot be used to measure the efficiency of a project because these expenses go on even if the project never existed. Therefore, they cannot be attributable to the life or the efficiency of the project or product. The last word has, obviously, to be written on the subject, but requires careful thought.

Life Cycle

The subject of life cycles was discussed earlier, and is assuming increasing importance. Goodman (1970) reported on a study of some 250 products over a 5-yr span. Two major conclusions resulted: First, there is a specific, recognizable, definable, life cycle for a product. The study concluded that on an average, a new product will last about 3 yr. This means that within that span a product must undergo its development, test marketing, national introduction, absorb the advertising and promotional funds spent, and, in addition, recoup a satisfactory return. All of this in 3 yr! This adds to our understanding of the growing focus of corporate managements on improving the success and shortening the period needed in development of new products.

Second, it is possible to regenerate a product at least once, and on an average; this regeneration might add another year to the life of a product. An average new product is thus expected to last for about 4 yr providing it continues to have support. Consider also that to create a successful new product, estimates indicate that it takes about 100 new product ideas to create 1 successful new product.

Another matter which should be borne in mind is that within the next 5-yr span, many believe that only about 20% of the sales at that time will come from products which exist today. The number of new product ideas which must be carefully screened and worked on to produce the one, successful new product suggests a future work load of some magnitude.

It behooves each of us to benefit as rapidly as we can from the experiences of others. The use of Fig. 10.1 developed by Goodman, as a decision guide for new product development, is one way to focus on the business decisions demanding attention.

OPERATIONS RESEARCH–FINANCIAL PROJECTIONS IN NEW PRODUCT DEVELOPMENT

Another possibility bears analysis. Although the example deals with a supplier of ingredients to mass market companies, the translation to our interests here is apparent as the example is unfolded.

The probability theory technique can be applied to such new business development decisions. Carson (1970) suggested selected steps that a decision maker can take leading to a final decision. The terms he uses are defined below.

Probability: The likelihood of an event occurring, based on the relative frequency of like events occurring in the past. If chances are 50-50 of something occurring, this would be expressed as a probability of 50%.

Expected Payoff: If we choose a particular course of action, how profitable it might become.

Prior Analysis: Decisions on the basis of past experience. This segment of Bayesian Theory is known as prior analysis.

Posterior Analysis: The reexamination of decisions made during prior analysis as a result of information added to the decision maker's experience.

Preposterior Analysis: The option of delaying a decision pending the receipt of more information.

He reported on a typical new product sequence to illustrate how the theory is applied. Assume a research laboratory has developed a new ingredient (a new product to them) which appears to be a major breakthrough. The product looks promising, although there are unanswered questions. The product manager using this system decides to examine the payoff at different sales levels by drawing a conditional payoff matrix, shown in Table 10.6

He determines that if the product sells well (provides sales of 10 million units throughout its product lifetime) it will contribute $5 million to earnings. If it has sales of 1 million units throughout its lifetime there will be a loss of $2 million. Should the product manager recommend building new facilities or should he recommend dropping the product?

TABLE 10.6

CONDITIONAL PAYOFF MATRIX (PROFIT IN MILLION DOLLARS)

Course of Action	Potential Sales Level	
	High (10 Million Pounds)	Low (1 Million Pounds)
Produce product	5	−2
Do not produce product	0	0

Source: Carson (1970).

TABLE 10.7

PRIOR ANALYSIS EXPECTED PAYOFF BASED ON PRESENT INFORMATION

Course of Action	Proba-bility of High Sales		Conditional Payoff (High Sales)		Proba-bility of Low Sales		Conditional Payoff (Low Sales)		Expected Payoff
Produce product	0.40	x	\$5 million	+	0.60	x	\$−2 million	=	\$0.8 million
Do not produce	0.40	x	0	+	0.60	x	0	=	0

Source: Carson (1970).

TABLE 10.8

POSTERIOR ANALYSIS REVISED PROBABILITIES OF SALES LEVELS

	Event		
	High Sales	Low Sales	Total
(1) Original probability	0.40	0.60	1.00
(2) Conditional probability of positive laboratory given event	0.75	0.20	
(3) Probability of both events occurring (1) x (2)	0.30	0.12	0.42
Revised probability of both sales levels given positive laboratory results	0.71	0.29	1.00

Source: Carson (1970).
[1] (3) adjusted to 1.00

Prior Analysis

Carson suggests this manager needs some estimate of product success. He considers: the success or failure of previous new products; the present economic situation; the past performance of the research group that developed the product; the morale of the sales force; or his own outlook; etc.

After considering all these, he may decide that the chances for high sales are about 40%. Thus, he has estimated that the probability for high sales is

40% and, conversely, that the probability for low sales is 60%. He has utilized conventional business judgement to assist in solving the problem (Table 10.7).

Next, he wants to know what the expected payoff would be if tested. The product manager also realizes that giving a sample to a potential customer increases the chance of a competitor prematurely learning of his new product. But his primary concern remains the high risk of having his product fail. Carson (1970) proposes he use posterior analysis to decide whether to have a customer laboratory test or to proceed without it, using the analysis shown in Table 10.8.

The product development manager sees that if the customer laboratory result is positive there is a 71% likelihood that the product will attain high sales. There is still a 29% chance of low sales in spite of a positive customer laboratory finding. This posterior analysis probability of 71% is much higher than his prior analysis probability, however.

Carson reports that next he should calculate what the new payoff would be with the new information acquired through the test, as shown in Table 10.9.

TABLE 10.9

POSTERIOR ANALYSIS PAYOFF TABLE WITH LABORATORY RESULTS

	Probability of High Sales		Payoff at High Sales		Probabiliby of Low Sales		Payoff at Low Sales		Expected Payoff
Produce product	0.71	X	$5 million	+	0.29	X	$−2 million	=	$3.0 million
Do not produce product	0.71	X	0	+	0.29	X	0	=	0

Source: Carson (1970).

Should he produce the product if the customer test result is positive? Carson's answer: clearly, yes. The expected payoff is $3.0 million as opposed to zero for not producing. The real question is whether or not to run the risk of a competitor analysis. As noted before, without the customer laboratory test, the expected payoff is only $0.8 million. The manager can increase his expected return $2.2 million by arranging for a test (provided results are positive). If the potential customers give a thumbs-down after the test, the expected payoff is $0.8 million if product is produced. The expected payoff is zero if the product is not produced.

The product manager decides the expected gain of $2.2 million is worth the risk of obtaining a wide consumer reaction, so he proceeds. The results are positive, and he hopes that a competitor will not bring out a new product like his own.

He feels confident now of writing a recommendation to spend $6 million to build a new plant and to initiate a program to sell the product.

Decisions are made by human beings. Carson suggests that we take a pessimistic outlook of possible events and viewpoints to illustrate one further option and its analysis. The product manager's report is written, his manager is away, and he awaits his return to make the recommendation. The day before his manager returns, the product manager develops some personal problems and wants no trouble at work. But, he is recommending an expenditure of $6 million. If the product fails to sell, the company will lose $2 million. And there is still a 29% likelihood based on present information (his original estimate plus the positive regional result) of a failure.

Preposterior Analysis

The product manager, Carson suggests, could recommend that R&D build a small test production plant. Output from this operation would be sufficient for a while. Then, if that goes well, he could recommend building a full scale plant. But this will take time and cost money. How long? How much? Is this course worth considering?

He learns that it would take 1 yr to build a test plant and would cost $800,000. Salaries of the technicians operating the plant for 1 yr would cost another $100,000. Total cost: $900,000. Nevertheless, this is much less than $6 million required for a new plant.

Returning to past records, he checks whether the first customer tests of previous products offered useful information about their future sales.

Table 10.10 shows that 8 out of 9 products which attained high sales were forecast by a positive customer test. This seems to be a good indicator of product success.

TABLE 10.10

FIRST CUSTOMER TEST RESULT RELATED TO EVENTUAL SALES RESULTS

	Previous Product Results	
Trial Run	High Sales	Low Sales
Positive	8	1
Negative	1	2
Total	9	3

Source: Carson (1970).

From this he deduces: Probability of positive result giving high sales is 8/9 = 0.88. Carson indicates that this can be calculated simply by multiplying the probability for high sales times the payoff for high sales, and adding the result to the value determined by multiplying the probability for low sales by the payoff for low sales. The accompanying Table 10.11 shows the results of this analysis by Carson.

TABLE 10.11

PREPOSTERIOR PROBABILITIES

	Event		
	High Sales	Low Sales	Total
(1) Original probability (prior analysis)	0.40	0.60	1.00
(2) Revised original probability after successful lab. test	0.71	0.29	
(3) Conditional probability of positive trial given event	0.88	0.33	
(4) Probability of both positive trial and event occurring (2) X (3)	0.62	0.10	0.72
(5) Rerevised probability of event occurring given positive trial run[1]	0.86	0.14	1.00

Source: Carson (1970).
[1]Adjusted to 1.00

Thus, if the product manager must decide what course of action to take, he knows that the expected payoff is greater if he recommends in favor of a factory for production. His decision should net $800,000 more profit than his alternative—not to produce.

However, the product manager is also aware that the company can ill afford to make a $2 million mistake. And that is what will happen if sales only reach 1 million units during the product's lifetime. He believes there is a 60% chance of that happening.

Posterior Analysis

Using Carson's example, assume that under similar circumstances in the past, this manager has had previous new products. He compares the results of these tests with the ultimate sales results obtained by those new products to determine whether the present test results are a good indicator of product success or failure.

Analysis (Table 10.12) reveals that of 24 high sales products, 18 were forecast by the first tests. Only twice in 10 times did findings show up positive for a product which subsequently generated low sales. The first regional test result appears to be an excellent forecaster of future sales success.

The manager next writes these conditional probabilities:

Probability for positive findings given high sales is

$$18/24 = 0.75$$

TABLE 10.12

FIRST CUSTOMER TEST RESULT RELATED TO EVENTUAL SALES RESULT

	Previous Product Results	
Lab. Results	High Sales	Low Sales
Positive	18	2
Negative	6	8
Total	24	10

Source: Carson (1970).

Probability for positive findings given low sales is

$$2/10 = 0.20$$

Probability of positive result given low sales is

$$1/3 = 0.33$$

What would a favorable trial run tell him? How much is it worth to have a run? He requires still another analysis, shown in Table 10.13.

TABLE 10.13

PREPOSTERIOR ANALYSIS EXPECTED PAYOFF WITH POSITIVE TRIAL RUN RESULTS

	Rerevised Probability of High Sales		Discounted Payoff		Rerevised Probability of Low Sales		Discounted Payoff		Expected Payoff
Produce	0.86	X	\$4.5 million	+	0.14	X	\$−1.8 million	=	\$3.6 million
Do not produce	0.14	X	0	+	0.14	X	0	=	0

Source: Carson (1970).

Carson points out that more information tends to raise confidence for high sales. Prior analysis showed 40%; posterior analysis showed 71%; and preposterior analysis 86%. The product manager now sees that if the trial run is favorable, there is an 86% likelihood of the product generating high sales nearly 9 times out of 10. What is the expected payoff?

Payoff must be discounted 1 yr since the revenues would now be produced 1 yr later. Table 10.13 shows this. The expected payoff if product is produced is \$3.6 million and zero if not.

However, the manager must subtract the \$900,000 incurred for test production plant and salaries; thus the expected payoff becomes:

Produce: \$3.6 million – \$0.9 million = \$2.7 million

Do not produce: \$0 – \$0.9 million = –\$0.9 million

But from Table 10.9 it is seen that the expected payoff after only a laboratory test was $3.0 million if product was produced.

Carson suggests that this tells him it is best to avoid the delay involved in building a pilot plant. His expected payoff is reduced $300,000 by this delay. He should recommend to commit $6 million to build a full scale plant.

This illustration taken from Carson shows Decision Theory in action. It incorporates business judgment, human frailties, and hard facts.

Financial control of research projects is essential because initial forecasts are seldom accurate. A good financial control system provides: (1) Methods for forecasting profitability as accurately as possible. (2) A system for rapid and simple updating of objectives. (3) A management system for the scientific personnel working on a project. (4) Enables the project's status to be easily appraised.

Most workers in the field would tend to agree that not enough attention is given early enough to the profitability consideration in new product development. The best advice available would be to adopt whatever techniques being found useful by others, and modify these to fit specific company situations. But, in any event, use some proven financial business guide.

Given a sound financial outlook for a new product development, there are further technical aspects which bear careful attention. The next chapter carries us into certain legal manifestations involved in new product inventions.

BIBLIOGRAPHY

Angelus, T. L. 1970. Improving the success ratio in new products. Food Technol.*24,* No. 4, 29-36.

Anon. 1970. Finance. Analysis of corporate earnings, stockholder relations, small business financing and new techniques. Harvard Business Rev. Reprint Serv., Soldier's Field, Boston.

Carson, G. D. 1970. Product development decisions using Bayesian decision theory. Food Prod. Develop. *2,* No. 6, 46-50.

Cavalier, P. A. 1967. The financial function in product development. Food Prod. Develop. *1,* No. 6, 28-30.

Cavalier, P. A. 1968. The Balance Sheet. Food Prod. Develop. *2,* No. 1, 32-34.

Goodman, S. R. 1970. Using decision guides for research and development. Food Technol. *24,* No. 4, 42-47.

Greyser, S. A. 1964. The case of unproductive products. Harvard Business Rev. *42,* No. 4, 38-47.

Sampson, R. T. 1969. Sense and sensitivity in pricing. Harvard Business Rev. *42,* No. 6, 45-49.

Scheuble, P. A. 1964. ROI for new product planning. Harvard Business Rev. *42,* No. 6, 118-128.

CHAPTER 11

Legal Aspects of New Products

Legal matters are matters pertaining to, based upon, or governed by law. The law is the binding custom or practice of a community and includes rules of conduct enforced by the controlling authority. That which is lawful is that which is in conformity with the principles and spirit of law, be it moral or juridical. That which is legal is that which is in conformity with the latter, or rules of the law, as administered. That which is legitimate is that which is sanctioned by known authority, or logically admissible, or, in conformity with acceptable standards. New product development must be all of these to be successful. In this chapter, we will explore some important legal and legitimate and lawful aspects of new product activities.

PROTECTING IDEAS AND SUGGESTIONS

Inside Sources

Sources of ideas and suggestions from inside a company need adequate consideration in order to encourage the staff at all levels to make what may be useful and profitable suggestions. This requires the establishment of a system for handling suggestions which will promote personnel awareness of company interest in their ideas which may have a bearing on the success and profitability of the enterprise. Various methods are used by different concerns to stimulate suggestions by their employees. If a company does not have an adequate system of this type in operation, its development may be worth considering. This requires the development and passing on to employees of information on the company policy in this area and must, of course, have careful legal consideration of the various aspects of company liability. Some provision for awards based on the value of adopted suggestions is certainly well worth attention.

Outside Sources

Suggestions from sources outside a company present problems which probably can best be solved by development of a company policy. A listing of a few of the problems arising from unsolicited suggestions from outside sources based on experience in mass marketing companies may be helpful. (1) Suggestions as to advertising themes or catch phrases. (2) Suggestions about a new product. Sometimes it may already be under development or projected. (3) Suppliers, particularly the smaller ones, may bring to R&D or to some other person or department within the company a product of undisclosed composition designed for a specific use. Each of these may, unless very carefully handled, result in undesirable legal actions against a company.

These situations and many others indicate the need for the development and implementation of a company policy on handling of suggestions of all types from sources inside and outside the company.

COMPANY PATENT RECORDS

Buckles (1957) found that there was no hard and fast rule as to the form of a written invention disclosure. Inventors have been known to make records on odd scraps of paper, shirt cardboards, or old paper bags. Such informality, however, is not encouraged for obvious reasons.

Permanent Notebooks Needed

A better practice, most believe, is to keep written records of all R&D work in bound notebooks. Most corporations provide their research workers with notebooks designed for this purpose and urge them to make daily entries. Loose-leaf or spirally-bound notebooks from which pages may be removed, or to which pages may be added, should not be used. A bound notebook having stitched-in pages is the preferred form of record. There can be no question at a later date of the possibility of pages having been inserted or removed.

With this type of notebook it may not be absolutely necessary to have every page signed and witnessed, but this practice is preferred. An important advantage of having a witness sign and date each page of the notebook entry, aside from enhancing the value of the witness' testimony years later, is that when this is done, it may be possible to introduce into evidence a photostat copy of a selected page, or only a few pages, without the necessity of disclosing the entire notebook, or even the entire entry relating to the particular experiment. Buckles (1957) found that often a patent interference or infringement case will involve only a single claim, and the supporting evidence for this specific part of an invention may be found on one page of the inventor's notebook. When this situation occurs, the inventor, or his company, usually does not want to show the opposing party more of the secret notebook contents than necessary. On the other hand, if each page has not been separately signed by the inventor and witnessed, it may be necessary to use as evidence more pages from the notebook than would otherwise be required or desired. He found that to save time and effort in having each page witnessed, laboratory personnel may be provided with a simple rubber stamp containing the following:

Recorded by ______________________ on _____________
(inventor) (date)

Witnessed and Understood on _________________________
(date)

By ______________________________________
(witness)

This can be stamped at the bottom of each notebook page. If management has special notebooks prepared for its employees' use, this notation may be printed directly on the bottom of each page when the notebook is manufactured.

The general practice is to have notebook pages consecutively numbered. If any pages or portions of a page are left blank, they should be crossed out by diagonal lines. For permanence of record, and to reduce the possibility of alteration, all entries should be made in ink or indelible pencil. If an error is made in any entry, legal advice available suggests that it should not be erased but should be crossed out. Where computations are necessary it is frequently a good plan to make all such calculations on a left-hand page, facing the page on which the written description or sketches are entered.

Some companies have found it advantageous to provide notebooks having alternate pages of transparent tissue, or tracing paper, on which a carbon copy of notebook entries may be made. If these tracing pages are perforated near the seam of the book they may then be readily removed for purposes of duplication, by blueprinting or otherwise, without destroying the original permanent record in the bound notebook. The use of a ball-point pen with this type of notebook is recommended, because it produces both a permanent record and good clear carbon copies for use elsewhere.

Notebooks of the recommended type should be kept by all researchers who are assigned to work on R&D projects. All notebook entries should be made by the individual in his own handwriting, and not by an assistant or secretary. If the researcher has assistance on a project, each assistant should be provided with a notebook of his own and should be instructed as to the proper manner of recording his thoughts and experiments.

LABORATORY NOTEBOOK CONTROL

An example of a laboratory notebook sheet and typical instructions for its use are shown in Fig. 11.1, 11.2, and 11.3.

The notebooks and sheets are numbered serially, as noted above. Instruction and index pages are bound in the front of each book. Additional instructions are usually contained in an operational or procedural manual.

Each notebook contains about 50 bound sheets. Each bound sheet is followed by 1 or 2 tear-out sheets with each of these "sets" of sheets printed with the same notebook and page numbers.

The notebooks, in practice, are generally issued by a laboratory administrative office. A serial log is kept there of the assignee, department, date of issue, supervisor, and date of return. Often a card file is kept on notebooks to facilitate audit of outstanding notebooks. This also becomes useful in personnel changes, etc.

Returned notebooks are stored in vaults or other safe places. The tear-out sheets themselves may be kept in the laboratory unit itself. The notebooks are also made available for loan within the laboratory on a need basis, as well as made available to patent and legal personnel, on the same basis.

(Company Name)

(Address)

BOOK NO. 750 **ASSIGNED TO** ____________

DATE STARTED ____________ **DATE FINISHED** ____________

INSTRUCTIONS

1. This notebook is the property of (Company) and must be returned to the office from which issued when completely filled. Until then it is in the care or the last person to whom it is assigned.

2. The attached index must be completed. No notebook will be accepted for filing if the index is incomplete. This index should be kept up to date.

3. All observations, notes and data arising out of or pertinent to your experimental work are to be entered directly in these notebook pages together with all comments, interpretations and predictions. All entries should be made in such fashion as to be perfectly understandable to anyone studying the pages and in sufficient detail to enable the experiments recorded to be duplicated.

4. Use the right hand pages for making entries. Preferably, entries should be made in ink, ball-point pen or indelible pencil. No erasures should be made and any unwanted entry should merely be crossed out but left legible.

5. All entries should be made consecutively leaving no blank pages. Later entries relating to a given experiment should identify the prior experiment by suitable cross-reference.

6. The date should be entered at the beginning of each day's work. If data pertaining to more than one experiment is entered on any one page, each separate entry should be adequately identified. Any unused portion should be ruled out.

7. Date and sign each page and have each page witnessed and dated by a co-worker familiar with the work you are doing.

8. Write clearly and keep your notes in good order. Your notebook is the original and, in many cases, the only authentic record of your work. Its value lies not only in the subject matter it contains but in the manner kept.

THIS BOOK CHECKED AND APPROVED BY ____________ (Department Head)

____________ (Date) FOR FILING.

FIG. 11.1. EXAMPLE OF LABORATORY NOTEBOOK INSTRUCTIONS

Each notebook should be labeled on the outside cover with the name of the individual whose notebook it is, the title of the project or department to which he is assigned, and the dates of starting and ending the work recorded.

Usually patents are assigned to the employer. It often happens that by the time suit is brought on a patent, the inventor may no longer be employed by the assignee. The inventor may have moved on to another company, his whereabouts unknown, or he may be deceased. In such cases, the evidence of the inventor's handwriting and signature as recorded in his notebook may be the only proof available (Buckles 1957).

Records of Teamwork

When two or more researchers work together on the same project, it is possible for them to keep a joint notebook; it is better, however, for each worker to keep a separate notebook.

Where each member of a research team may contribute partial solutions to an overall problem, the project leader should record the combined development in his notebook, with notations identifying the individuals who are responsible for the separate parts. In this way it will be possible to determine at a later date who contributed what to the project.

In group effort, it is probable that all will be coinventors of the combination, a problem with which the patent attorney must struggle.

The regular and systematic keeping of laboratory notebooks also affords a chronological record of development activity which may be used to prove diligence. It is important that all experiments should be recorded, including the failures as well as the successes, for the unsuccessful experiments may show a continuing effort toward the ultimate successful completion of the invention.

PATENT DISCLOSURES

Some procedure should be established for periodic review of notebook entries, with a view toward identifying possible inventions that have been made, and initiating patent applications. This notebook review may be made by the patent attorney, or patent coordinator, who may be hired for that purpose. Someone in close touch with the overall problem to which the research is directed may have a broad perspective that will enable him to perceive inventive contributions, perhaps even more readily than does the close-up detailed viewpoint of the researcher who actually makes the contribution, Buckles suggests.

However, if researchers are encouraged to be "patent conscious" (as they should be) and if they know the state of the prior art in which they are working, then they should be able to recognize when they may have made an invention, and they should take the initiative in calling their disclosure to the attention of the laboratory management or the patent department, whichever system is used in a company.

(Company Name)

(Address)

LABORATORY NOTEBOOK INSTRUCTIONS

This notebook is the property of (Company)

All information recorded in it will be regarded as confidential. It should be treated as official property of the Corporation and must not be taken from the Research Center without supervisory approval. While it is in your hands, it is your responsibility and you must take every precaution to safeguard against loss. In case of fire, theft or disappearance of this book, your supervisor should be notified in writing. The Technical Files Section has assigned this notebook to you on a charge basis. When all pages are used, you should return it promptly to the Technical Files Section. If you have need for it in the future you may obtain it on a loan basis.

All notebook entries should be made in ink, using a ball point pen. Do not use pencil.

Insert clean carbon paper when needed to insure legible duplicates.

Project title, project number and date should be recorded on each page.

The carbon copies of all pages (duplicates) are retained permanently in the binder; the original notes (made on the perforated pages) are removed and filed in the appropriate project folder.

Do not make erasures in this notebook. Delete unwanted data simply by drawing a straight line horizontally through the data.

Unused portion of page should be indicated by two crossing diagonal lines in the unused area.

Make all entries consecutively, leaving no blank pages.

When a page or a day's entry is completed, sign your name at lower right. Have your signature witnessed at the lower left. This witnessing means that the page was filed at the time of signing.

Pages are provided for a Table of Contents. This should be completed to enable ready access to the data in the future.

IMPORTANCE OF NOTEBOOK:

Original entries systematically and properly made in the usual course of your work are the best record of your accomplishments. This notebook may be the deciding factor in litigation involving a question of inventorship. It may also be the only complete source of information for future experiments, or for the preparation of reports you may be required to write, THEREFORE:

1. As part of your daily routine, enter clearly and concisely all your experimental data, plans for experiments, analysis of results, calculations, observations or ideas. These data shall be placed in the notebook immediately and directly. Do not use looseleaf books or loose pieces of paper for this purpose.

2. As part of your weekly routine, prepare a weekly summary in the notebook for supervisory review.

TO MAKE THIS BOOK MOST USEFUL:

Plan your presentation before writing so it can be easily read and understood.

State clearly and completely the procedure followed, giving all conditions of the experiment (temperatures, pressures, time required, etc.) and apparatus used (include sketches if necessary). This notebook should provide sufficient data to enable any of your associates to duplicate the procedures and results.

Give all results and observations including references to literature, analytical data, and pertinent calculations. Data may be presented in tabular form and in graphs to great advantage. Cross sectional ruling is provided for your convenience in aligning tables and making grpahs and sketches.

Promptly disclose any new and possibly patentable ideas occurring thereon to one of your associates who is familiar with the work you are doing (preferably the Section Head or one who has actually seen the experiment carried out), and have him affix his signature and the date. He should write in "disclosed to" at the time of signing.

When an idea, process, etc., of any importance is finally found to be workable, the process, etc., should be demonstrated in the presence of witnesses, other than co-inventors, who are able to understand the process. These witnesses should sign the page in the notebook describing the demonstration carried out in their presence, writing in "demonstrated to and understood by". Before this procedure is followed, the Section Head in charge of the project should be notified.

FIG. 11.2. EXAMPLE OF LABORATORY NOTEBOOK INSTRUCTIONS

Continued From: (Company Name) **N B 750**

N B____________ (Address) **P G 4**

P G____________ Date________

SUBJECT OF EXPT.

Signed:______________________Date:_______

Experiment seen and understood

by me:______________________Date:_______

FIG. 11.3. EXAMPLE OF LABORATORY NOTEBOOK WORK SHEETS

When the engineer or scientist has reason to believe that he may have made an invention, or if he simply suspects that his contribution may possibly be inventive, then he should make a separate written disclosure. This disclosure should ordinarily be clear in presentation, witnessed, and submitted for review of the patent attorney.

INNOVATIONS AND THE U.S. FOOD AND DRUG ADMINISTRATION (FDA)

There are a number of legal aspects involved in introducing new food products to the market. They are becoming more pressing with each passing day. They tend to frustrate the mass marketing food industry.

Can new concepts be found in interpreting the provisions of the U.S. Food, Drug, and Cosmetic Act, and similar laws in other countries, which will accelerate the progress being made in developing new food products?

Depew (1969) believed that such new concepts can be found without jeopardizing the protection to which the consumer is entitled. This is the subject of this section. While the focus will be on U.S. laws, similar situations also exist in other countries. Truly imaginative efforts will be needed if the public interest is to be served in the long term.

THE BASIC LAW

According to Depew, the most important legal provisions of the U.S. Food, Drug, and Cosmetic Act of 1938, relative to their impact on new food product development, are Sections 409, as implemented by Section 402 (a) (2) (C) covering food additives; 401, as implemented by Section 403 (g) covering food standards; 403 (c) covering imitations; and 403 (j) covering special dietary foods.

From the record accumulated under Sections 401 and 403 (g) of the Act, it is readily apparent that one of the foremost purposes of the Congress was to prevent the debasing and/or cheapening of food products by way of economic adulteration. Section 401 provides, in part, as follows:

"Whenever in the judgment of the Secretary such action will promote honesty and fair dealing in the interest of consumers, he shall promulgate regulations fixing and establishing for any food, under its common or usual name so far as practicable, a reasonable definition and standard of identity, a reasonable standard of quality and/or reasonable standards of fill of containers . . . In prescribing a definition and standard of identity for any food or class of food in which optional ingredients are permitted, the Secretary shall, for the purpose of promoting honesty and fair dealing in the interest of consumers, designate the optional ingredients which shall be named on the label."

Section 403 (g) provides that a food shall be misbranded: "(g) If it purports to be or is represented as a food for which a definition and standard of identity has been prescribed by regulations as provided by Section 401, unless (1) it conforms to such definition and standard, and (2) its label bears the name of the food specified in the definition and standard, and, insofar as may be required by such regulations, the common names of optional ingredients (other than spices, flavoring, and coloring) present in such food."

A large number of foods have been standardized since this law went into effect. The ingredients permitted in standardized foods have been specified in detail. Such standards are commonly called recipe standards. It was determined in a count test in 1943 that in the language of the Supreme Court, the statutory purpose to fix a definition of identity of an article of food sold under its common or usual name would be defeated if producers were free to add ingredients, however wholesome, which are not within the definition (Oser 1968).

It is worth noting that individual producers or processors, the FDA itself, consumer groups, or trade associations may initiate steps toward the establishment of a food standard, according to Depew's analysis.

If FDA concurs, or if FDA initiates steps toward a new or revised standard, it publishes a proposed standard in the Federal Register, the official government publication. At the same time, it invites written comments, including objections, before a specified date, Oser reports.

Prior to publication of a proposed standard, the FDA may consult with the National Academy of Sciences, trade associations, and others.

The greatest possible latitude for free discussion of the scientific problems involved exists at that time. Scientists from universities, research centers, commercial and industry laboratories have the opportunity to be heard. In this atmosphere, differences are usually reconciled. In establishing standards, the judgment of scientific experts prevails. This should be the consensus of experts in the particular field (Depew 1969).

As soon as practicable after completion of hearings, the FDA enters an order which is published in the Federal Register. The order is based on substantial evidence of record and sets forth detailed findings of fact on which the order is based. In case of controversy as to the validity of an order, persons adversely affected may file a petition for review with the U.S. Court of Appeals.

Provisions governing the holding of hearings and appeals thereof are Section 701 of the U.S. Food, Drug, and Cosmetic Act and Sections 552-558, 701-706 of the Administrative Procedures Act.

However, after years of research, a food manufacturer may discover that a substance may advantageously be used in a standardized product. In order to use it, according to Depew, one must disclose exactly what the substance is and show why its use will improve the standardized product. If he can establish that, the standard can be modified to permit use of the new ingredient. At the same time, all competitors may also reap the benefits of his research unless the ingredient or the process of incorporating it in a standardized product is protected by patent.

If the manufacturer is unable to convince FDA that his substance in standardized products promotes "honesty and fair dealing in the interest of consumers," he will then have wasted much time and money in making his discovery. The standard will exclude it.

For these and other reasons, people support the view that food standards should deal with the basic integrity of food products, and not be rigid recipe standards. Depew (1969) stated that any ingredient not specifically forbidden by the standard should be allowed, if properly declared on the label, as long as the basic integrity of the product is maintained, is safe, and has the characteristics expected of it. The situation is presently being challenged by consumer groups for the opposite reason, i.e., too much has been permitted to be added to foods. The entire matter is unresolved to date and will not be easily resolved.

In this connection, the Joint FAO/WHO Codex Alimentarius Commission has drawn a set of Guiding Principles which includes: "Unless clearly necessary, avoid 'recipe' standards, i.e., those which exclude the use of other than specified ingredients." The protocol for standards provides for composition and essential quality factors with the object of preventing fraud.

MISBRANDED PRODUCTS

Section 403 (c) of the U.S. Food, Drug, and Cosmetic Act provides that a food shall be deemed to be misbranded: "If it is an imitation of another food, unless its label bears, in type of uniform size and prominence, the word 'imitation' and, immediately thereafter, the name of the food imitated."

Lawyers suggest this provision was directed at what was regarded in the 1930's as economic adulteration. The word imitation was selected as a term commonly used to differentiate such a product from the genuine, and was intended to be construed within its dictionary definition, according to Depew. Dictionary definition of the word is "simulating something superior." The difficulty currently is that the development of modern technology has produced new foods that are superior to foods which they seem to resemble. What FDA may consider an imitation may in fact be functionally superior. "Imitation" salad dressings with reduced fat content may well be considered by overweight Americans, human physiologists, and the medical profession to be superior to regular salad dressings; or, a more expensive dietary meat food product may be made by a new process lowering the salt content. FDA has recognized this dilemma and is seeking solutions. The best advice available in such matters is to consult them at an early date in the development cycle.

NEW PRODUCTS

Section 403 (j) is also pertinent to those who may wish to innovate in terms of the nutritional value of their products. It provides that, "if a food purports to be or is represented for special dietary use, it shall be misbranded unless so labeled as to bear such information concerning its vitamin, mineral, and other dietary properties as the FDA determines and by regulation prescribes as necessary in order to fully inform purchasers as to its value for such use." FDA's position is that enrichment above that correcting deficiencies is wasteful and contrary to the interest of most consumers, and tends to confuse consumers as to their nutritional needs. Again, this matter is due for revision in the near future due to rising pressures from consumer groups, Depew suggests.

Section 409 provides that a food additive, as defined in Section 201 (s) is outlawed from interstate commerce unless its actual or intended use conforms with an effective administrative regulation issued prescribing the conditions under which it may be safely used.

A great deal could be said about food additives. Much has been said most effectively by FDA representatives and industry representatives. Reference to current literature is most desirable for trends.

Many companies were started with a new idea which others shared. It behooves us all to bring about prudent interpretations of the law which will encourage future pioneers to bring discoveries to the consumer.

Suffice it to say that there is a rising tide of consumerism which will not easily be discounted by those desiring success in new product development.

In the next chapter, we will explore further the business milieu and consumerism, as we examine the business of change and the impact of change on business.

BIBLIOGRAPHY

Anon. 1970. Business policy. Making and content of policies. Harvard Business Rev. Reprint Serv., Soldier's Field, Boston.

Anon. 1970. Ethics for Executives. In marketing, public relations, promotion, business policy. Harvard Business Rev. Reprint Serv., Soldier's Field, Boston.

Buckles, R. 1957. Ideas, Inventions and Patents. John Wiley & Sons, New York.

Depew, F. M. 1969. Legal aspects of food innovations. Food Prod. Develop. *3,* No. 2, 42-48.

Hunter, M. W. 1969. Are technological upheavals inevitable? Harvard Business Rev. *47,* No. 5, 73-83.

Kleinfeld, V. A. 1969. Nutrition, food standards and the FDA. Food Prod. Develop. *3,* No. 2, 36-40.

Nagel, A. H. 1970. The impact of food standards on food marketing. Food Prod. Develop. *4,* No. 2, 92-96.

North, P. Q., and Pyke, D. L. 1969. Probes of the technological future. Harvard Business Rev. *47,* No. 3, 68-82.

Oser, B. L. 1968. Where is industry's voice in food regulations. Food Prod. Develop. *2,* No. 2, 40-42.

SECTION III

The Business Milieu

CHAPTER 12

Merchandising and New Products

It is on the supermarket shelf that the new product must finally find a place of its own. In moving products from production lines to supermarkets, a host of issues appears, not the least of which is the impact of new products and their interference with the already established systems and spaces in supermarkets. To understand how these are handled, it is believed useful to explore that which is happening in merchandising, then look at how a new product enters the system. What is the battle of supermarket shelf space like?

BIRTH OF THE SUPERMARKET

During the decade following World War I the food industry settled down to a period of steady growth, increasing total sales before the crash in 1929. During this same period the chain store system of food merchandising was also growing. This was a major subject for discussion. At the time, the food chains were damned by some and seen as a mixed blessing by others. Few could perceive what a tremendous part they were to play in the future growth of the industry.

Important technological advances affecting the industry came along with startling rapidity during the 1920's and 1930's. One such advance was the development of bulk storage and handling of ingredients. This development has been perfected to the point that today many companies in the industry utilize computer-directed control systems which automatically deliver the exact quantity of each ingredient required directly from storage to mixer (Cahn 1969).

MOISTUREPROOF CELLOPHANE

In the early 1930's, during the depths of the depression, moistureproof cellophane was introduced for packaging. This material proved to be a great boon to the industry. Its transparency, combined with its excellent protective qualities and the fact that it could be easily and effectively heat-sealed, made it exceptionally desirable for packaging. It enabled the industry to package its products with good protection of the contents, yet so that their goodness could actually be seen within the packages when they were displayed in the self-service food store (Cahn 1969).

SUPERMARKET REPLACES CORNER STORE

The industry's bulk selling-cracker barrel era was coming to a close. With the introduction of moistureproof cellophane the floodgates were opened to an onrush of new packaging ideas. Up until this time, most of the industry's packages consisted of closed cartons overwrapped with opaque printed waxed paper. Cartons with window openings and various types of trays overwrapped with cellophane, made their appearance. Then came bags made of printed cellophane leading to the development of today's high-speed machines which automatically, and at high speeds, form, fill, and seal bags of virtually every size and type. New materials and methods have been introduced almost continuously during recent years. The industry's packaging methods have undergone constant evolution, always with the consumer's interest in mind, Cahn (1969) found.

Packaging improvements have gone hand in hand with the growth of the self-service supermarket in America. The vast difference between the old box and barrel days and the packages of foods now to be seen on the shelves of a typical supermarket are evidence of the changing scene.

COMPUTERIZED MERCHANDISING

In the food industry, manufacturers are geared to supply supermarkets on a high volume basis with turnover and profits at a very high level. Haase and Seawell (1970) found that manpower, costs, and profits were naturally geared to this "fast in and out" method of business. The advent of the quick stop-and-shop limited purchase type of buying has brought added cost of doing business due to the rising costs of sales personnel, delivery, and service on a per order basis, they found.

Maximum sales and profits are needed by both the retailer and the manufacturer. Both must be partners in serving consumers and generating profits.

Haase and Seawell (1970) report that there has always been a tendency in the grocery business to stock as many varieties as possible in an attempt to keep Mrs. Consumer from going down the street to shop. They also found that the selectivity of brand and variety is extremely important. It may not always be profitable to stock several lines or brands due to slower turnover and less physical space.

A common problem is that of the "three too's" of the supermarket industry—too many suppliers, too much space, too much inventory. Supermarket space management is an important management tool, they believe. If retailers could select the proper amount of space for each commodity category, then successfully manage the commodities within this space, maximum sales and profits would result at all times.

These two segments of retailing are, of course, difficult to achieve. How Nabisco approached the situation was described by Haase and Seawell (1970) wherein they described their company's computerized space management program. The essential factors in this program for national supermarket chains are:

(1) Total sales in study period. (2) Department sales. (3) Special displays. (4) Turnover of all merchandise. (5) Total linear feet of shelving. (6) Number of varieties and packs and totals by companies. (7) Average department sales per row. (8) Department sales per linear foot of shelving. (9) Average retail inventory value per linear foot of shelving. (10) Actual stockturns per month (department only). (11) Percentage of total inventory by companies to sales by companies. (12) Sales to space by companies. (13) Print out of leading sellers. (14) Print out of categories, by variety and company. (15) Recommendation on deletions. (16) Complete marketing analysis by products. (17) Complete space allocation proposal on space management findings. (18) A new return on investment program. (19) A guide on future planning of packaged food departments based on volume, and many other factors which may apply to your own situation.

This led to the development of the "mini-survey" to accommodate the smaller supermarkets and convenience stores. This study is quick, accurate, and can be managed by local sales people and store management, Haase and Seawell report. Their program is presented to each store. It starts with a completely blank form (Form 1, Fig. 12.1).

Form 2 shows the present space allocations by companies. This is filled in by the company's sales personnel and is verified by the store manager. After the study is completed, a recommendation is made on space allocation on actual sales in dollars to each company.

Though these mini-studies are not nearly as sophisticated as the computerized space management studies, a revealing analysis of all products can be made.

STRENGTHS AND WEAKNESSES

Haase and Seawell (1970) also reported on 2 computerized studies (A and B) conducted in 2 store chains, 4 stores to each study. They believed that both studies, being typical and average, pointed out and revealed strengths and weaknesses in the store departments analyzed.

Table 12.1 shows space-to-sales percentages in Study A.

Table 12.2 shows the space-to-sales results for Study B.

TURNOVER BY SUPPLIER

Table 12.3 from their studies, indicates the turnover of products by suppliers, all of whom had less than 1 turnover per 4-week period.

Table 12.4 shows similar results for the department from Study B. One supplier was getting less than 1 turnover every 4 weeks, and the other less than 1 turnover every 8 weeks. Next, they analyzed the top 6 varieties in a category, the 6 varieties that move out of the store the quickest, regardless of supplier.

SALES AS PERCENTAGE OF SPACE

In analyzing the top 6 varieties in their Study A (refer to Table 12.5), they

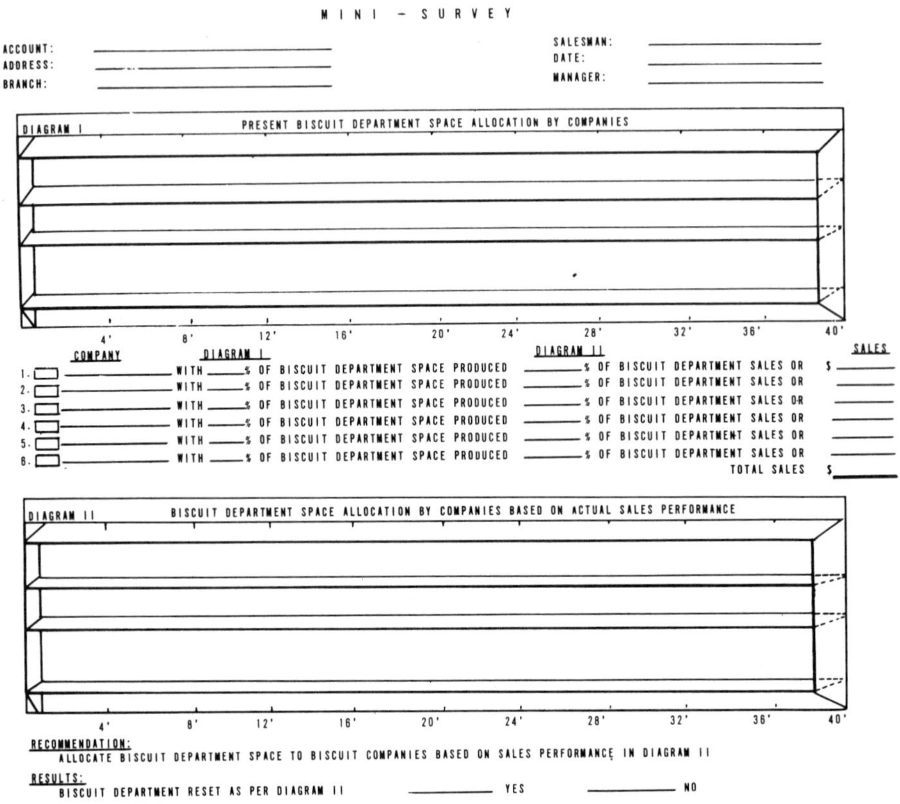

MINI - SURVEY

ACCOUNT: ____________
ADDRESS: ____________
BRANCH: ____________

SALESMAN: ____________
DATE: ____________
MANAGER: ____________

DIAGRAM I — PRESENT BISCUIT DEPARTMENT SPACE ALLOCATION BY COMPANIES

4' 8' 12' 16' 20' 24' 28' 32' 36' 40'

COMPANY	DIAGRAM I	DIAGRAM II	SALES
1. ____	WITH ____% OF BISCUIT DEPARTMENT SPACE PRODUCED	____% OF BISCUIT DEPARTMENT SALES OR	$____
2. ____	WITH ____% OF BISCUIT DEPARTMENT SPACE PRODUCED	____% OF BISCUIT DEPARTMENT SALES OR	____
3. ____	WITH ____% OF BISCUIT DEPARTMENT SPACE PRODUCED	____% OF BISCUIT DEPARTMENT SALES OR	____
4. ____	WITH ____% OF BISCUIT DEPARTMENT SPACE PRODUCED	____% OF BISCUIT DEPARTMENT SALES OR	____
5. ____	WITH ____% OF BISCUIT DEPARTMENT SPACE PRODUCED	____% OF BISCUIT DEPARTMENT SALES OR	____
6. ____	WITH ____% OF BISCUIT DEPARTMENT SPACE PRODUCED	____% OF BISCUIT DEPARTMENT SALES OR	____
		TOTAL SALES	$____

DIAGRAM II — BISCUIT DEPARTMENT SPACE ALLOCATION BY COMPANIES BASED ON ACTUAL SALES PERFORMANCE

4' 8' 12' 16' 20' 24' 28' 32' 36' 40'

RECOMMENDATION:
ALLOCATE BISCUIT DEPARTMENT SPACE TO BISCUIT COMPANIES BASED ON SALES PERFORMANCE IN DIAGRAM II

RESULTS:
BISCUIT DEPARTMENT RESET AS PER DIAGRAM II ____ YES ____ NO

Courtesy of Haase and Seawell (1970)

FIG. 12.1. MINI-SURVEY COMPUTER SPACE MANAGEMENT FORM

found 182 varieties on the shelf which took up 472 rows of space and produced total sales of $1290. The top 6 varieties took up 36 rows and sold at $253 for 20% of the sales. To put it concisely, 3% of the varieties produced 20% of the sales in 8% of the space. The No. 1 variety produced $74 during this 4 week period and the No. 6 variety, $26.

In the same situation taken from their Study B (refer to Table 12.6), 112 varieties in 332 rows produced $588 in sales in 4 weeks. They found the 6 leading varieties produced sales of $170 out of 38 rows and accounted for 29% of the total sales. In other words, 5% of the varieties produced 29% of the sales in 11% of the space. In this study, the No. 1 variety produced $46 and the No. 6 variety produced $17.

Volume varieties from the slowest movers are relevant for the sake of comparison. In Study A, they found, the weakest 61 varieties amounted to 34% of the varieties stocked, took up 118 rows or 25% of the total rows, and produced

TABLE 12.1

PERCENTAGE OF SPACE TO SALES IN STUDY A

Company	Percentage of Space	Percentage of Sales
X	38	48
Y	38	34
Z	24	18

Source: Haase and Seawell (1970).

TABLE 12.2

PERCENTAGE OF SPACE TO SALES IN STUDY B

Company	Percentage of Space	Percentage of Sales
X	50	65
Y	50	35

Source: Haase and Seawell (1970).

TABLE 12.3

TURNOVER IN STUDY A

Company	4-Week Turnover
X	0.8
Y	0.5
Z	0.7
Total	0.7

Source: Haase and Seawell (1970).

TABLE 12.4

TURNOVER IN STUDY B

Company	4-Week Turnover
X	0.8
Y	0.4
Total	0.6

Source: Haase and Seawell (1970).

\$102 in sales. Haase and Seawell (1970) found that this was 8% of the total sales, and averaged 57¢ per row for the 4-week period, or 14¢ per row per week. The average unit price in this study was 42¢. Sales of these varieties amounted to

TABLE 12.5

SIX VOLUME VARIETIES ANALYZED IN STUDY A

Total Varieties	Total Rows	Total Sales $	Sales of Top 6 Varieties $	Total Rows of Top 6	Percentage to Total Sales of Top 6
182	472	1290	253	36	20

Source: Haase and Seawell (1970).
Note: 3% of the varieties produced 20% of the sales in 8% of the space.

TABLE 12.6

SIX VOLUME VARIETIES ANALYZED IN STUDY B

Total Varieties	Total Rows	Total Sales ($)	Sales of Top 6 ($)	Total Rows of Top 6	Percentage to Total Sales of Top 6
112	332	588	170	38	29

Source: Haase and Seawell (1970).
Note: 5% of the varieties produced 29% of the sales in 11% of the space.

TABLE 12.7

COMPUTERIZED RECOMMENDATION FOR STUDY A

Company	Deletions	Linear Feet Present Shelf Space	Linear Feet Proposed Shelf Space
X	15	112	69
Y	19	111	40
Z	17	70	35
Total	51	293	144

Source: Haase and Seawell (1970).
Note: Space reduction of 149 linear feet in 4 stores is 37 linear feet per store, or about a 9-ft section.

$^1/_3$ of a package per week. If the inventory averaged 6 packages per variety, there would be an 18-week supply on the shelf.

In the 4 stores they surveyed in Study B, the weakest 18 varieties amounted to 16% of the total varieties stocked, took up 34 rows for 10% of the total, and produced $10 in sales. This was 2% of the total sales, and the average sale per row per period was 29¢, or 7¢ a week.

COMPUTERIZED

Their computer recommended (Table 12.7) that 51 varieties be deleted because they were not profitable. It also recommended a reduction in space from 293 linear feet to 144 linear feet, in order to boost turnover to their recommended 1 turn per month, or, from about an 18-ft section to a 9-ft section.

Their recommendations for Study B, (refer to Table 12.8) included deletion of 15 varieties (by name) and reduction of shelf space from 192 linear feet to 99 linear feet. Again, these figures were for 4 stores, so what the computer really recommended in Study B was to reduce the departments from approximately a 12-ft section to a 6-ft section.

TABLE 12.8

COMPUTERIZED RECOMMENDATION FOR STUDY B

Company	Deletions	Linear Feet Present Shelf Space	Linear Feet Proposed Shelf Space
X	7	96	64
Y	8	96	35
Total	15	192	99

Source: Haase and Seawell (1970).

They cautioned that while not all stores need a reduction in space, some stores have such volume that keeping merchandise on the shelf is a problem. The computer pointed this out, too.

They summarized the following necessary elements that help make a supermarket department as productive as it can be:

Turnover

Turnover is the key to all profits. Turnover of all merchandise in a category must be brought up to at least one turn per month.

Selectivity

Selectivity not only on varieties stocked, but also by companies. Studies have proven that a very few varieties do the bulk of the volume and that even these cannot stand too much duplication. It is important to use many categories as a guide to reach this selectivity.

Space

Space should be allocated to companies on a volume basis. But the total amount of space allocated to a certain (any) department should be limited by the selectivity of product and by the total gross volume of the store.

Manufacturers have a challenge from retailers. That is, bring the retailer all the tools they possess to help guide departments in stores to "maximum sales and profits." The computerized approach points the way to the future.

INFLUENCE OF ADVERTISING ON SUPERMARKET BUYERS

To touch briefly on the impact of advertising on sales, Table 12.9 indicates the reaction of buyers for large retail stores to supplier advertising. Morrill (1970) found, as is well known, that advertising has a drawing effect, not only on consumers, but also on supermarket chain operations; and, if a series of such chains do not stock a new product, its potential is nil. Consumers will not have the opportunity to see the product in these markets, and these markets do a lion's share of the retail food business.

LOGISTICS OF DISTRIBUTION

Integration of new products into the flow of goods from plant to markets bears careful attention. In order to capitalize on this opportunity, management must: (1) Think of the physical product distribution process as a system in which all the components and functions must be properly balanced. (2) Take a fresh look at the responsibilities, capabilities, and organization of traffic, warehouse management, inventory control, and other functions which make up the overall system. (3) Reexamine the company's physical plant and distribution procedures in the light of technical advances in such areas as transportation, data processing, and materials handling. (4) Develop new products which are compatible with the system (Magee, 1960).

Let us first examine the pressing need for improved distribution systems, then outline some of the most promising ways by which progress in industrial logistics can be achieved with special attention to the implications of technological advances, the problems getting started, and the steps that should be taken to ensure good distribution.

Magee (1960) found the need for progress in distribution is the result of not one but several trends in costs, in product-line policy, and in the market place. More often than not, the challenge posed is to the system as a whole, not just to the particular part or function where trouble is most obvious.

While for years businessmen may have looked with mixed feelings on the increase in distribution costs in our economy, over the past half century Magee found that tremendous strides may have been made in reducing the costs of production, but these strides have not been duplicated in all other areas. If the overall efficiency of companies is to continue to improve, management must turn its attention increasingly to holding distribution costs in line. He found physical distribution costs, in particular—estimated by some to represent the third largest component in the total cost of doing business—are a logical center for attention.

In many production operations it has been possible in the past to substitute a machine for human labor and to cut the cost of one operation without seriously disturbing the rest of the production system. This is not the case in efforts to cut physical distribution costs. Indiscriminate cost reduction in any one of the individual cost elements, such as inventory maintenance, warehousing, transportation, or clerical activities, Magee found, can have a disastrous effect on the

TABLE 12.9

OPINIONS OF BUYERS ABOUT THREE COMPETING SUPPLIERS WHEN EXPOSED AND UNEXPOSED TO THEIR ADVERTISING

	Proportion of Buyers Holding These Opinions					
	About Company A Which Ran 17 Pages of Advertising in 1 Yr.			About Companies B and C Which Did Negligible Advertising in 1 Yr.		
Opinions	Exposed	Unexposed	Differential (%)	Exposed	Unexposed	Differential (%)
Preferred as supplier	0.37[1]	0.30	+23	0.09	0.12	−25
Second and third choice as supplier	0.19	0.19	…[2]	0.14	0.14	…[2]
Willing to consider	0.56	0.50	+12	0.24	0.28	−14
Leads in quality	0.20	0.18	+11	0.03	0.03	…[2]
Leads in price	0.05	0.07	−29	0.03	0.03	…[2]
Leads in delivery	0.18	0:15	+20	0.04	0.04	…[2]
Best in your past experience	0.22	0.18	+22	0.05	0.05	…[2]
Best technical assistance	0.20	0.14	+43	0.04	0.04	…[2]
Salesmen lead in product knowledge	0.13	0.08	+63	0.05	0.05	…[2]
Salesmen lead in service	0.11	0.09	+22	0.02	0.02	…[2]
Most enthusiastic salesmen	0.12	0.06	+100	0.02	0.02	…[2]
Share of market						
In customers	0.56	0.47	+19	0.20	0.23	−13
In dollars	0.38	0.35	+9	0.13	0.19	−32

Source: Morrill (1970).
1 Read: 37% of the buyers exposed to A's advertising preferred A as supplier.
2 No significant difference.

efficiency of the system as a whole. Obviously, a new product must fit into an existing system if effective distribution is to be gained at reasonable cost.

One might cut inventories. Certainly a reduction in inventories will save capital investment and the costs of supplying capital. It may also save some expenses in storage, taxes, and insurance. On the other hand, an indiscriminate reduction in inventory levels may seriously impair the availability of products in the field. An inventory reduction which saves money but destroys competitive position is hardly a contribution.

One might cut transportation costs by shipping in larger quantities and taking advantage of volume carload or truckload rates. But if lower transportation costs are achieved at the expense of slower or less frequent movement of goods, there are risks in cutting the flexibility and responsiveness of the distribution system to changes in customer requirements. This might require greater field inventories and create greater investment requirements and hence risks, he found.

Proliferating Product Lines

Physical distribution systems in recent years have been put under tremendous pressure induced by changes in product-line characteristics. Until recently, for example, products were largely utilitarian with differences in product characteristics rather closely related to function.

Today, an almost unending list of products in the consumer field with slight variations in color, packaging, and other features have imposed heavy burdens on distribution systems. In the marketing of goods, variations in grade and size have had similar impact. The wide variety of package sizes required for consumer products has led manufacturers to demand correspondingly wider ranges of packaging materials from suppliers, and these demands have created difficult problems of scheduling and inventory control, according to Magee.

The growth and change in product-line characteristics in consumer products have meant that manufacturing plants have had more items to make, and the distribution system has had more items to handle and to stock. More items mean lower volume per item and correspondingly higher unit handling inventory and storage costs. Take, for example, just the impact on inventory requirements of substituting 3 items for 1.

Magee's studies (1960) found that if we substituted items B, C, and D for an old item, A, and if sales among these items would break down 60% to B, 30% to C, and 10% to D, with no overall increase in sales compared to volume on the old item A, then field inventory requirements would increase by more than 60%. This figure is based on characteristic relationships between inventory and sales (see Table 12.10). In general, the larger the sales the lower the inventory can be relative to sales. Thus, product D with 10% of sales needs a much higher proportion of inventory than product B with 60% of the sales. At a carrying cost of 20% a year, this increase represents a very high expense for maintaining such a competitive position, Magee found.

TABLE 12.10

AN EXAMPLE OF THE FRACTION OF TOTAL SALES ACCOUNTED FOR BY THE FRACTION OF TOTAL ITEMS IN THE PRODUCT LINE

Percent of Items in Production Line	Percent of Sales
10	70
20	84
30	90
40	93
50	95
60	96
70	97
80	98
90	99
100	99

Source: Magee (1960).

However, if we assume that items, B, C, and D do more than yield the same total volume, say a total volume increased by 50%, then the inventory requirements would double, and inventory cost per unit sold would increase over 30%–a substantial source of pressure on the distribution system, Magee found.

These figures illustrate the impact of small-volume items on the cost of operating the distribution system. Yet diversity of product sales is characteristic in business. The typical relationship between the number of items sold and the proportion of sales they account for, is usually based on the records of a large number of firms in the consumer products fields. Table 12.11 reveals that while 10-20% of total items sold characteristically yield 80% of the sales, $^1/_2$ of the items in the line account for less than 4% of the sales. It is the bottom $^1/_2$ of the product line that imposes a great deal of difficulty, expense, and investment on the distribution system, according to Magee.

Increased costs and product-line pressures suggest that management should take a hard look at alternative distribution patterns, as a means of cutting logistics costs without a major sacrifice in service.

Magee suggests that a company can carry only central stocks of low-selling items.

For many items in the line, a good compromise may be to carry some low- or middle-volume items in only a few large regional warehouses, as a compromise between excessive storage costs incurred from broad-scale stocking and the transportation and service penalties incurred by attempting to meet demand from manufacturing points alone.

Warehouse points can be consolidated. With increased volume through the individual warehouses, carrying a broader product line at the local points begins to make greater economic sense.

The first and most basic job of the distribution system is to service orders and to help transform them into sales. As business grows more competitive and the

TABLE 12.11

AN EXAMPLE OF WHAT HAPPENS TO INVENTORIES WHEN THE PRODUCT LINE IS BROADENED

Number of Items in Line	Items	Percent Sales	Percent Inventory	Total Percent Inventory
1	A	100		
	A		100	100
3 (With same total sales)	B	60		
	C	30		
	D	10		
	B		75	
	C		55	
	D		30	160
3 (With 50% increase in sales)	B	90		
	C	45		
	D	15		150
	B		95	
	C		65	
	D		40	200

Source: Magee (1960).

public has become harder to please, management has focused increasing attention on the quality of its logistical operations. What can be done to make new and existing products more readily available for purchase in local markets? What improvements can be made in backing up new product merchandising and advertising programs with adequate deliveries and service?

In analyzing the capacity of a distribution system to produce sales, it is useful to examine three key characteristics, according to Magee:

(1) Location. It has been estimated that with 5 distribution points a company can reach 33% of the U.S. consumer market within a day; from 25 warehouse locations, 80% can be reached in 1 day.

(2) Inventories. Approximately 80% more inventory is needed to fill 95% of the customers' orders out of stock in a typical business than to fill only 80%.

(3) Responsiveness. The effectiveness with which distribution transmits needs back to the supplying plant and gets material needed into the stores determines how quickly the business can shift with changes in customer preferences to meet demand with minimum investment and cost.

To sum up, a number of pressures are focused on today's distribution systems. As manufacturing efficiency has increased and product cost has come down, distribution costs have gone up. Physical distribution costs are a significant share of these.

Business in many fields is becoming increasingly competitive, with competition taking new forms in availability of goods and reliabilty of delivery. Magee identified product changes which are forcing new pressures on the distribution system—more items to carry, faster obsolescence and inventory turnover. In particular, changes in merchandising practices, as discussed earlier, have significantly complicated the distribution problem. Pressures for improvement in distribution logistics also include internal forces such as the need to stabilize production and to insulate production levels from short-term fluctuations in sales. It is obvious now that the choice of distribution system each company makes will have a significant impact on product design, plant investment, and organization. The distribution system also will have significant influence on new product development in the formulation of corporate plans for the decade ahead.

INTRODUCING THE NEW PRODUCT

Based upon all the evidence available suggesting that a new product is ready for selling in stores, a plan emerges concerning the specific area in the country to "roll out" the product. The system used will depend upon the nature of the distribution organization a company uses. If it sells its own products, it can make plans to identify an area or division of the country to initiate the process. If it sells through brokers, it will take another direction. But in general, it all boils down to selecting some one area to introduce a new product. (It is rare in the food industry to initially sell a product nationally.)

This calls for development of plans to add a product to the existing sales organization program. If a company has divisional organizations around the country, then one division might introduce one product, and another a second, etc.

Working closely with sales organizations, it then becomes possible to introduce the product. The advertising program will then be initiated, and the new product starts its life struggle. If it is successful, then the new product will be "rolled out" into the next division, and on across the country.

Using the merchandising techniques discussed above, it soon will become clear what the result will be for all the effort and funds committed to the new enterprise. Success depends on teamwork and filling consumer needs, with profit for the retailer and the manufacturer.

In the next chapters, we will try to penetrate more deeply into the subjects of business growth and change, with emphasis on success for the enterprise.

BIBLIOGRAPHY

Anon. 1970. Advertising management. How to plan, organize and measure results. Harvard Business Rev. Reprint Serv., Soldier's Field, Boston.

Anon. 1970. Selling and sales management. Factual data from the field. Harvard Business Rev. Reprint Serv., Soldier's Field, Boston.

Anon. 1970. Production Management. Directing and planning operations. Harvard Business Rev. Reprint Serv., Soldier's Field, Boston.

Brown, R. G. 1959. Statistical Forecasting for Inventory Control. McGraw Hill Book Co., New York.

Cahn, W. 1969. Out of the Cracker Barrel. Simon and Schuster, New York.

Ditcher, E. 1962. The world customer. Harvard Business Rev. *40*, No. 4, 151-160.

Haase, A. E., and Seawell, E. R. 1970. Merchandising. Nabisco space management program. Snack Foods *59,* No. 2, 48-52.

Magee, J. F. 1960. The logistics of distribution. Harvard Business Rev. *38,* No. 4, 82-94.

Morrill, J. E. 1970. Industrial advertising pays off. Harvard Business Rev. *48,* No. 2, 4-14, 159-168.

Tallman, G. B., and Blomstrom, B. 1962. Retail innovations challenge manufacturers. Harvard Business Rev. *40,* No. 5, 162-127.

Wittreich, W. J. 1962. Misunderstanding the retailer. Harvard Business Rev. *40,* No. 3, 128-136.

CHAPTER 13

Improving Success Potentials of New Products

The reasons for failure in new product development in the food industry have received a good deal of attention. While no one study more than another has uncovered the answer, there is accumulated literature describing dozens of rather spectacular failures, either in test market, or at regional or national introduction levels. Such failures have happened in organizations and to people who thought they knew the system to use and followed all the rules. In the food field, the investigations covered well-known and some not-as-well-known corporate entries. However, these studies contain in their findings, some clues and possible insights on how to improve the success potential of the system used for new products. These will be explored in this chapter.

WHAT IS THE SYSTEMS APPROACH?

McDaniel (1964) found that the systems approach sprang to life during World War II. It was associated with how to bomb targets more effectively, with the Manhattan Project, and with studies of optimum search patterns for use in locating U-boats during the Battle of the North Atlantic. It reached its heights with NASA space exploration activities.

Adler (1967) suggests, as an aid to a decision-maker, that he choose a course of action by systematically investigating his proper objectives; one might compare quantitatively where possible the costs, effectiveness, and risks associated with the alternative policies or strategies for achieving them, and then formulate additional alternatives if those examined are found wanting.

The systems approach is useful in dealing with complex problems of choice that have multiple and possibly conflicting objectives. Put more broadly, the systems approach attempts to apply the "scientific method" to complex marketing problems studied as a whole; it seeks to discipline according to Adler (1967).

He reports that disciplining marketing is no easy matter and that marketing must be perceived as a process rather than as a series of isolated, discrete actions; competitors must be viewed as components of each marketer's own system. The process involves a flow and counterflow of information through marketing communications, as in the cycle of "listen (i.e. marketing research), publish (messages, media), listen (more marketing research), revise, publish, listen . . ." Bauer (1963) found the desires of the communicator alone are but part of the picture.

Levitt (1965) identified the interactions between five different forces—source effect (i.e., the reputation or credibility of the sponsor of the message), sleeper effect (the declining influence of source credibility with the passage of time); message effect (the character and quality of the message); communicator effect (the impact of the transmitter, e.g., a salesman); and audience effect (the competence and responsibility of the audience). Adler (1967) reported efforts to model the entire purchasing process, and to make mathematical models of the entire marketing process.

MARKETING INTELLIGENCE CENTER

Adler (1967) saw examples of corporate application of the systems approach. Figure 13.1 is a schematic portrayal of these relationships. He found that the systems approach was effective in new product management. If the purpose of business is to create a customer at a profit, then the needs of the customer must be carefully served. Adler believes that the customer buys nutrition (not bread), beauty (not cosmetics), and warmth (not fuel oil), etc. Some disagree.

At this point, we believe that most will agree that the pressure to originate successful new products cannot be met without method. It must be calculated to enhance the probabilities of profitable innovation. Many believe that the tradi-

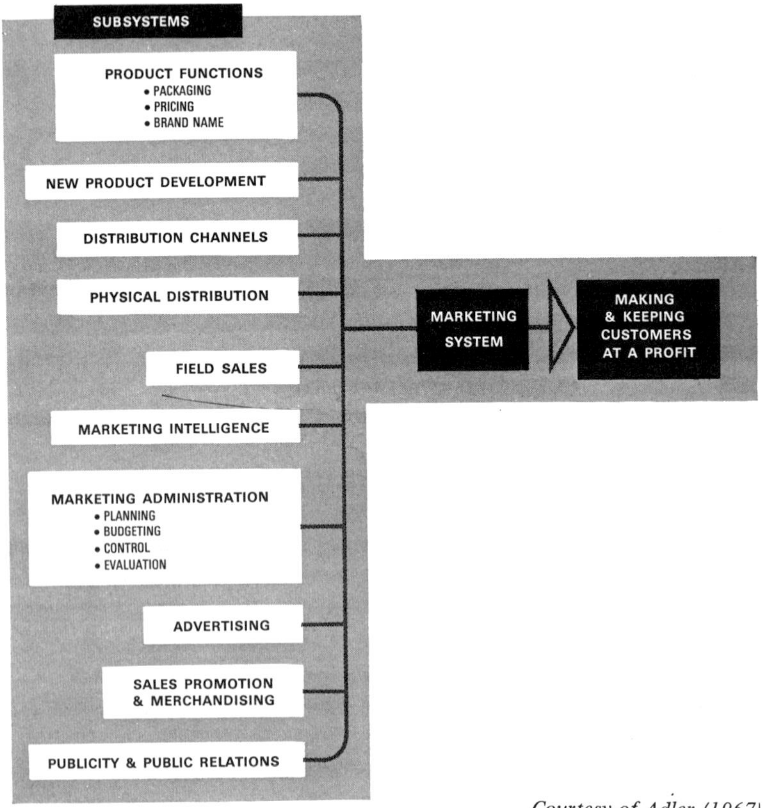

Courtesy of Adler (1967)

FIG. 13.1. MARKETING SUBSYSTEMS AND THE TOTAL SYSTEM

tional notion of marketing research is becoming antiquated. It has tended to concentrate on the study of tiny fractions of a product problem rather than on the problem as a whole.

The role of the systems approach is to help evolve a marketing intelligence system tailored to the needs of each company. Such a system would serve as the nerve center, according to Adler, with these major characteristics (see Fig. 13.2): (1) Continuous surveillance of the market. (2) A group of research techniques used together. (3) A network of information sources. (4) Integrated analysis of data developed. (5) Concentration on reporting practical, action-oriented recommendations.

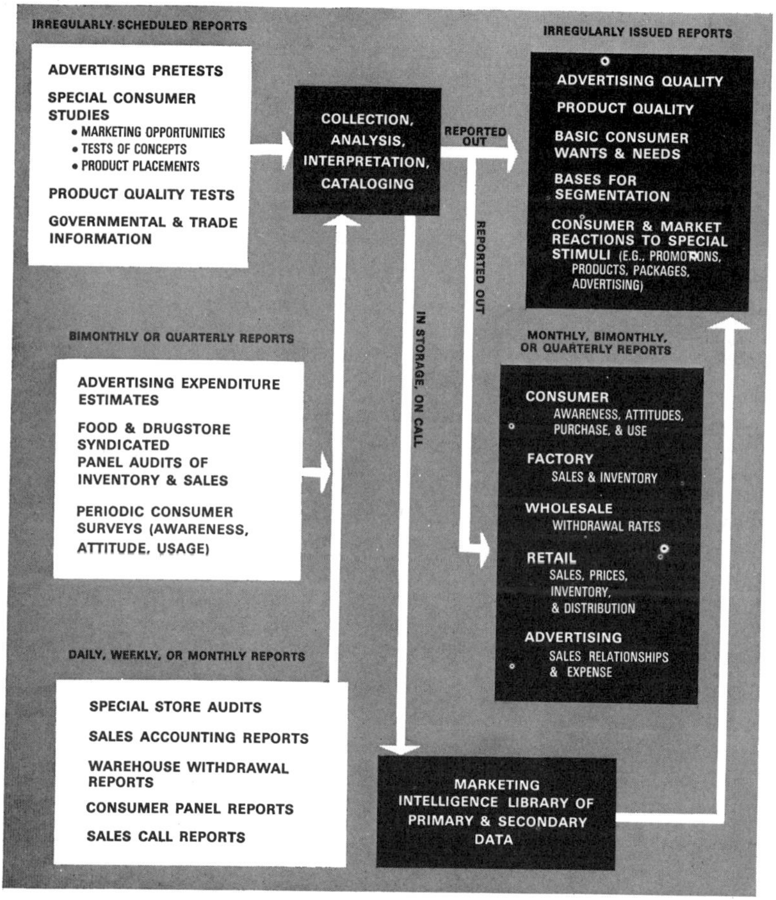

Courtesy of Adler (1967)

FIG. 13.2. MARKETING INTELLIGENCE SYSTEM

Concept in Use

A practical instance of the use of such an intelligence system was shown by Adler (1967). It provides information from these sources: (1) A continuing large-scale consumer market study covering attitudinal and behavioral data dealing with weight control. (2) Nielsen store audit data, on a bimonthly basis. (3) A monthly sales audit conducted among a panel of 100 high-volume food stores in 20 markets to provide advance indications of brand share shifts. (4) Supermarket warehouse withdrawal figures from Time, Inc.'s new service, Selling Areas-Marketing, Inc. (5) Salesmen's weekly reports which, in addition to serving the purposes of sales management control, call for reconnaissance on competitive promotions, new product launches, price changes, etc. (6) Advertising expenditure data, by media class, from the company's accounting department. (7) Figures on sales and related topics from company factories. (8) Competitive advertising expenditure and exposure data, supplied by the division's advertising agencies at periodic intervals. (9) A panel of consumers.

Adler believes such a systems approach permits a company, with the help of analyses of factory sales data, warehouse withdrawal information, and consumer purchases from Nielsen, to monitor transactions at each stage of the flow of goods through the distribution channel and to detect accumulations or developing shortages. He found that management was also able to spot sources of potential problems in time to deal with them effectively. For example, if factory shipments exceed consumer purchases, more promotional pressure is required. Conversely, if factory production lags behind consumer purchases, production effort must be increased.

NEW PRODUCT ADVERTISING AGENCY VIEW

Yet in spite of all the best efforts, new product failures continue to occur at rather surprising rates. While the systems approach tends to reduce risks, it would be useful to uncover more about why new products are failing.

Diagnostic research by Angelus (1970) revealed that high on the list of most frequent failure characteristics involved the inability of the consumer to distinguish significant differences or advantages in the new form of product offered. The new products studied included use of frozen food instead of canned, aerosols instead of liquids, liquids instead of tablets, and tablets instead of powders.

Angelus (1970) and co-workers found that the proliferation of the brands of these products with inadequate advertising support were clues for failure—for reasons other than product form—as revealed in his diagnostic studies of consumer reactions.

Positioning problems, according to Angelus, fall into one of three areas—insignificant, confused, or mismatched.

He found insignificant positioning is highly overrated and, in most cases, is a direct reflection of insignificant product differences. It follows that if a manufacturer develops an insignificant product, the advertising will reflect this.

Generally, when the product difference is real to the consumer, advertising will fall into place naturally, he reports. Usually, the most straightforward execution of the difference is the most effective. Examples of real consumer differences with straightforward advertising are known.

Next, he finds that mismatched positioning is the greatest single problem in advertising because it is prevalent in both new and established products. Mismatched positioning results in media inefficiencies. The challenge in positioning a new product, he believes, is to determine the most effective combination of consumer appeal and product performance. A mismatch occurs when the most effective combination has not been found.

Mismatched positioning also results, his research reveals, when a product is developed with marginal product differences. The advertising generally takes the direction of the highest potential appeal area which may not relate to the performance characteristics of the particular product. In most product categories, experts say there are at least 100 combinations of appeal and performance. Against 10 appeal areas (and there probably are 10 more), there are some 10 to 15 product performance characteristics to choose from. All told, Angelus finds that there are 100 to 150 possible positioning combinations to choose from to achieve the most effective and efficient proposition.

Angelus found that "Me-tooism" may be an overworked term in the marketing man's vocabulary because 80% of all new products have a point of difference. "Me-too" products can be highly successful in certain situations, such as line extensions where a "me-too" can trade on a corporate brand name.

National brand quality images readily lend themselves to an additional line extension.

Experience reveals that "me-too" products can be highly successful in a market which is not saturated. Advertising and promotional pressures then become the determining wedge in building a franchise.

Market explosion involving the introduction of several new products of the same appeal is also the cause of many new product failures. Many manufacturers rush into such a situation, as was discussed in Chap. 4, hoping to gain a foothold. A spending spiral occurs and the smaller manufacturer gets squeezed out.

DeNisco (1970) found that in food products, because of the difficulty in assessing consumers' taste preferences in other than prolonged testing situations, performance problems occur.

Manufacturers should not delude themselves into believing that a point of difference will outweigh a product problem. A product test may not reveal a product problem, because the unique aspects of the product are artificially highlighted during the test.

Most observers of the scene believe that true product problems seem to be occurring with less frequency each year. More sophisticated laboratory development and interpretation of product tests and greater emphasis on prolonged market testing give the manufacturer a greater chance of uncovering and solving product problems, as we discussed in Chap. 5, and which will be presented in the next chapters.

Countless numbers of small manufacturers are experiencing repeated failures in entering new markets alien to their existing product line. According to Angelus and others, these failures come about because the manufacturers often overestimate their production, marketing, or financial capability.

ADVERTISING AGENCY ROLE

New products are not solely a new combination of ingredients. New products are a combination of research, manufacturing, packaging, marketing, and advertising. It is, therefore, important that each becomes an active part of the team if the effort is to succeed.

Science can play an important role in many of the creative operations of an advertising agency, according to DeNisco (1970). Some agencies employ qualified scientists in a science department consisting of registered pharmacists and medical consultants. When necessary, specialists—physicians, dentists, chemists, physicists, and others depending on the product and problem—are consulted.

Their knowledge and understanding of the product is the keynote to good, effective advertising. The main function of such a science department, according to DeNisco, is to provide technical information in layman's language to the writers, account executives, and lawyers giving information on: the product's uses, composition and function of each ingredient, how the product compares to competition, and what general areas of claims and possible demonstrations can be documented. With this basic information, advertising account executives can intelligently discuss the product with their client; writers can start creating; and lawyers can determine whether advertising claims can be scientifically supported.

Fortified with the facts, advertising writers have two alternatives—disregard them entirely or create within the framework of the known parameters. In either case, DeNisco reports that the validity and accuracy of advertising claims, assuming they involve product attributes, have to go through the science department of the agency.

Documenting Demonstrations

An important part of an advertising campaign, particularly for TV, is the visual presentation of the verbal selling proposition which is a demonstration of the selling point. Usually, the time allotted for a demonstration in a commercial is about 10 sec, according to DeNisco. It has to be a simple, valid, reproducible demonstration. No shortcuts or mock-ups are permitted by the FTC; fortunately, they are permitted to use time wipes and cuts to compress the time. It is not widely realized that each commercial involving a demonstration is documented in writing, and covers the following: studio at which the commercial was filmed; name and code number of the commercial; date; materials used in the demonstration; procedure; and results. Results are reported for each take. DeNisco notes that these reports are submitted to the networks in support of the validity of the demonstration as produced.

Legal Approval

Before any advertising is produced, he continues, either for TV or print, it has to be approved by the legal department. The science department must be able to document technical claims and demonstrations to the satisfaction of the legal department. Agency lawyers act as a miniature in-house FTC and FDA combined.

Performance Testing

The science department in an advertising agency may or may not have a laboratory. If it does, this laboratory is used primarily for applied research. According to DeNisco, they evaluate the clients' product as a consumer would and compare its performance to competitive products. In conducting these performance tests, they often come up with copy ideas and demonstrations for advertisements. They also try to identify product weaknesses and suggest means of strengthening the product's performance.

SUGGESTIONS FOR SUCCESS

Some conclusions have been reached by experts in the field as a result of study of new product failures. There are major problems on both the manufacturing side and the advertising agency side which are causes of a high mortality rate in new products.

Angelus (1970) lists these suggestions for success: (1) Distinguish between marketing opportunities and new product opportunities. (2) Create FULL new brand concepts. (3) Restructure and restaff new product department. (4) Use greater creative interpretation of market research. (5) Demand new approaches by advertising agencies.

Some believe that a basic issue in new product work is that manufacturers have not distinguished between marketing opportunities and new product opportunities. Marketing opportunities exist, according to Angelus and others, when certain market conditions make a particular market open for product expansion. Some of these conditions are summarized by Angelus: (1) Insignificant differences between major brands. A new product with a minor innovation might represent a meaningful point of difference to the consumer. (2) High profit returns. The established brands have become complacent. (3) Lack of innovations. Consumers may be receptive to a new product entry. (4) Brand switching. Low consumer loyalty to existing brands may turn out to be a real market opportunity.

What most failures reveal is that a new product opportunity exists only by going one step further—a creative step.

Must Develop Brand Concept Early

Most specialists in the field recommend that before any technical formulation or packaging development is started, how to sell the product must be determined. Full brand concepts contain name, package design, and appeal to motivate a consumer to try and use it. After a brand concept has been refined to the point where it is evident that a new product opportunity exists, product specifications and packaging development can begin.

Much unnecessary effort can be eliminated if the product specifications relate only to the brand concept. The time-consuming process of developing product and package design can be made more effective if these are focused at the start to fit definitive brand objectives—not a vague concept.

Restructure and Restaff

Experts in the field tend to agree that new products departments in companies may need to be restructured and restaffed to improve new product thinking.

Most people agree that the new products man should be willing to take a chance, willing to convince others of his belief, and willing to press ahead. He should not be second-guessed every step of the way, either.

In organizing the company new products department there is value in considering two teams—a conceptual team and an execution team headed by the type of individual described above. The execution team should be given responsibility for getting projects through to their successful conclusion expeditiously. Another route is to organize venture teams, as will be discussed later.

Reevaluate Market Research

The entire function of marketing research in new product development should be reevaluated. Many companies tend to rely too heavily on market research and consequently, have found an easy out for making decisions (Jenkins 1970).

Most market research in the new product area, except for simple concept evaluations and test market audits, is unactionable. Experts generally conclude that there is something wrong with present product test methods, analysis, and interpretation if the majority had successful product tests and failed. In short, what is needed is more sophisticated techniques used by more expert people.

New Advertising Agency Approaches

There is evidence to indicate that advertising agencies are often ineffective with new product assignments because they may be overly-structured to perform specialized functions needed in new product work, particularly for smaller companies.

Because of the present attitude of advertising agencies toward new products, one of the most important developments in the 1970's, Angelus feels, will be the growth of special new products agencies.

The new products agency will provide a different new product service. First, he forecasts, their attitude must be positive. Their income derives from profitable new product fees, with incentive bonuses if the new brand meets sales and profit objectives.

Second, there will be few departments, as such, in the new products agency—just teams. Each team will probably be composed of a creative marketing head with both agency and manufacturing experience, a concept manager, and a research specialist. This group will steer a project from concept development through market introduction. The new products agency will have package designers, R&D personnel, and promotion specialists who are expert in their fields.

The functions of the new products agency, he believes, will be to uncover new product opportunities, create and evaluate new brand concepts, create and screen new brand names, design packages and labels, offer technical guidance on product development, and specialize in developing introductory marketing plans. They will try to keep the creation of the new product (or brand) under one roof so that it emerges with a unified, cohesive, and singular brand image.

The new products agency, according to Angelus, will take the initiative by showing smaller manufacturers where real new opportunities exist and then creating new brands to exploit these opportunities.

Certainly, change of some kind is needed when, each year, more new products fail than succeed, and fail after being successfully market researched. The last word has yet to be written, obviously.

BIBLIOGRAPHY

Adler, L. 1967. Systems approach to marketing. Harvard Business Rev. *45,* No. 3, 105-118.

Angelus, T. L. 1970. Improving the success ratio in new products. Food Technol. *24,* No. 4, 24-36.

Bauer, R. A. 1963. Communications as a transaction. Public Opinion Quart. *Spring,* 83.

Cooper, A. C. 1966. Small companies can pioneer new products. Harvard Business Rev. *44,* No. 5, 162-179.

Crockett, J. G. 1969. The new products manager in product development. Food Technol. *23,* No. 6, 26-30.

DeNisco, S. 1970. Agency approach to new product development. Food Technol. *24,* No. 4, 22-24.

Harper, M. Jr. 1963. *Cited by* L. Adler. 1967. Systems approach to marketing. Harvard Business Rev. *45,* No. 3, 105-118.

Hilton, P. 1967. New Product Development. Prentice-Hall, Englewood Cliffs, N.J.

Jenkins, F. A. 1970. The culmination of product development and market research in consumer communications. Food Technol. *24,* No. 4, 60-67.

Levitt, T. 1965. Industrial purchasing behavior. Harvard Business School, Div. Res., Boston.

Levitt, T. 1966. Innovative imitation. Harvard Business Rev. *44,* No. 5, 63-70.

McDaniel, G. 1964. The meaning of the systems movement to the acceleration and direction of the American economy. Proc. 1964 Systems Engr. Conf., Clapp and Poliak, Inc., New York.

O'Meara, W. H. 1961. Selecting profitable products. Harvard Business Rev. *39,* No. 1, 95-102.

Oxenfeldt, A. R. 1966. Product line pricing. Harvard Business Rev. *44,* No. 4, 137-144.

Silk, L. S. 1960. The Research Revolution. McGraw Hill Book Co., New York.

CHAPTER 14

Market Exploration and Acquisitions

In today's economic climate, a company must grow to keep abreast of changes in a profitable way. One of the features of this growth is that a corporation has some real control over its destiny. Market exploration is useful in terms of identifying potential growth and expansion areas. Forecasting, as discussed in Chap. 3, is a preliminary activity. Next, an existing product market often can be segmented. Characterization of a corporation is obviously needed at the outset. Market segmentation and product positioning are also needed to develop meaningful long-range plans for internal growth. Combined with this, there exists the possibility of expansion through acquisitions and mergers. In this chapter, we will explore some available techniques in exploration of markets and product positioning, and follow this with guidelines for acquisitions and mergers.

MARKET EXPLORATION

A major salt producer needed a long-range plan for growth and expansion. Based only on an understanding of the company and its needs, Diehl (1969), in his original review, encountered an unusually clear example of market segmentation and product positioning. The specific product available was common table salt. He pursued the analysis for the following reasons: (1) The product seemed to capsule the history and present performance of the company. (2) It was the largest profit contributor in the corporation. (3) The product seemed to represent the company to all classes of customers and had the greatest consumer impact. (4) Among consumers, this particular product seemed central to their general idea of the company.

He found most consumers did not think about the actual product in its physical state, but they tended to figure out how many ways it could be used and what its actual value was. Such is the case, as shown in Fig. 14.1, for table salt.

TABLE USE	COOKING/ BAKING USE
MEDICINAL USE	HOUSEHOLD USE

Courtesy of Diehl (1969)

FIG. 14.1. AREAS OF USE OF TABLE SALT

When the customer is deciding on the product's usefulness, he not only thinks of the things that the product will do for him personally, but also impersonally.

Therefore, all uses for a particular product can be systematized by a four-cell grid, such as shown in Fig. 14.2.

	PERSONAL	IMPERSONAL
FOOD		
NON-FOOD		

Courtesy of Diehl (1969)

FIG. 14.2. SYSTEMATIZING THE USES OF TABLE SALT

Being limited to packaging, communication, and pricing, Diehl's research showed that cooking/baking (food/impersonal) was by far the largest use area for table salt.

Each practical use of table salt was then studied fully to see whether there were any marketing possibilities. An interesting discovery was made when, basically, a mixture of ingredients (salt plus another substance) was used to modify the product (table salt) and expand one of the consumer use areas. Other examples became evident, as shown in Fig. 14.3.

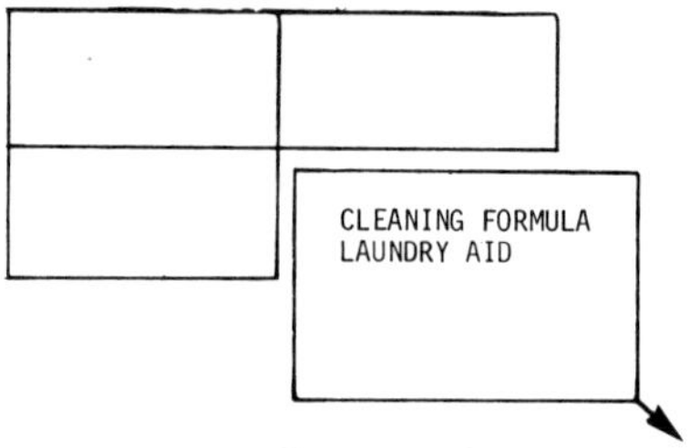

Courtesy of Diehl (1969)

FIG. 14.3. HOUSEHOLD USES OF TABLE SALT

Similarly, the other nonfood use area could be expanded, such as in Fig. 14.4.

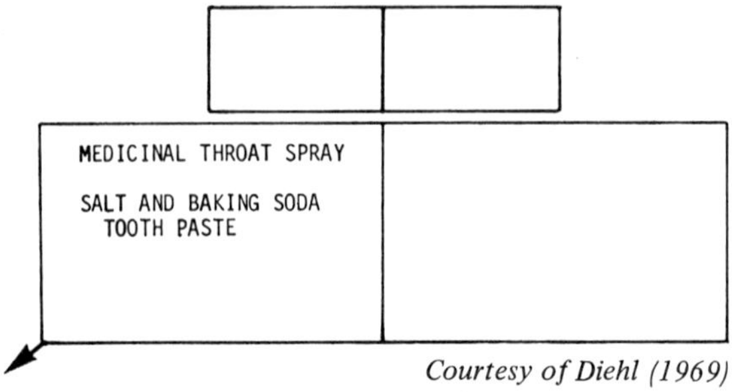

Courtesy of Diehl (1969)

FIG. 14.4. MEDICINAL USES OF TABLE SALT

He found that the same expansion could take place in the table use area, as in Fig. 14.5.

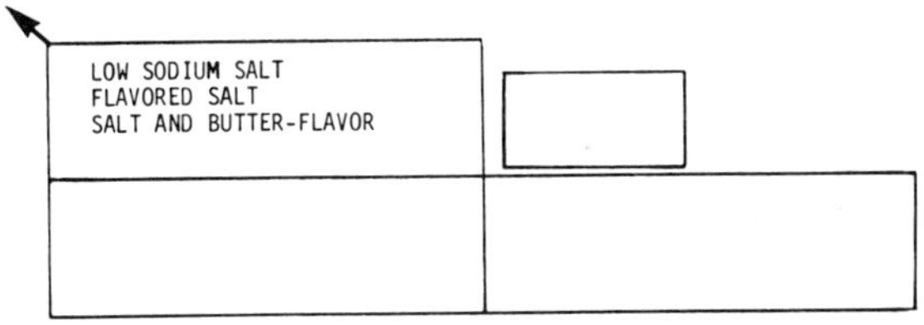

Courtesy of Diehl (1969)

FIG. 14.5. TABLE USES OF TABLE SALT

Also, the food/impersonal use area could be expanded to become a "line" of cooking/baking products (Fig. 14.6).

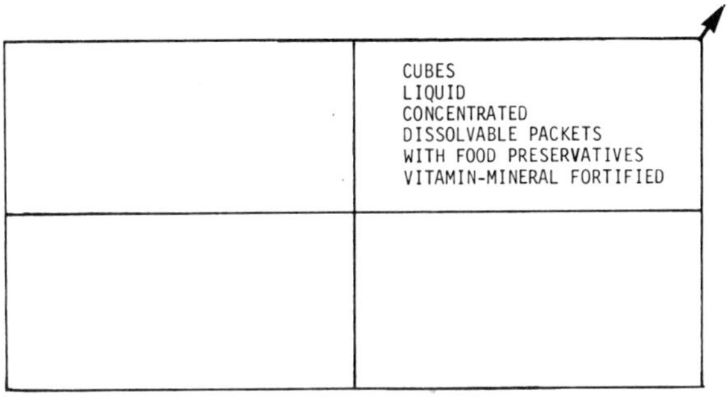

Courtesy of Diehl (1969)

FIG. 14.6. COOKING/BAKING USES OF TABLE SALT

Next, Diehl showed the results of market segmentation and product positioning in terms of the company itself. Market segmentation analysis showed the way and product positioning provided the means for exploiting the overall, long-term potential of the company. From his illustration below, still further growth could come by taking each expanded use area to its fullest potential (Fig. 14.7).

However, Diehl (1969) stressed that orderly growth should start at the center and that priorities should be given to the use areas. This brought out the point at which many business analysis presentations fail . . . the implicit or explicit question: Where to start for fullest growth potential? He felt the analysis led to a simple operational answer. He drew two conclusions from this exploration: First, opportunities arise for single products to be positioned more closely against clear market segments. Second, once a company has a successful product, the company should use this product market to its greatest potential and in as many different areas as is financially rewarding.

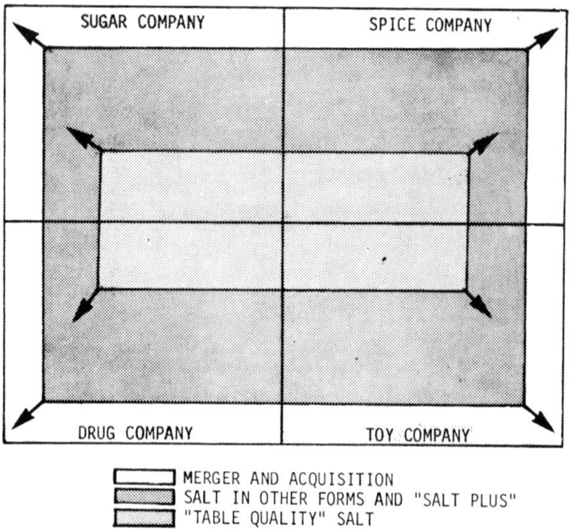

Courtesy of Diehl (1969)

FIG. 14.7. MARKET SEGMENTATION AND COMPANY EXPANSION POTENTIALS

ACQUISITIONS

Applications of the principles described above can be found in the diversification or acquisition plans of companies. Before proceeding in this direction, it is useful to discuss mergers and acquisitions in overall view, then proceed with specifics.

Planning for Mergers

MacDougal and Malek (1970) found two common obstacles that prevented acquisition from being successfully negotiated. One was the inability to settle on a realistic price. The other was the failure to be convincing on the nonfinancial aspects of the merger. These snags can be overcome, they suggested, by first analyzing and then graphically portraying the key relationships of various financial alternatives. Each should be able to grasp the trade-offs involved and the limitations of various price levels and terms. This approach, combined with a plan for the growth of the combined businesses, should result in a successful merger.

Mergers and acquisitions, they found, loom large in most companies' strategic growth plans. The pressures to acquire, whether to improve earnings and stability through diversification, or to fend off possible take-overs—preoccupy chief executives as well as staff planners. This is reported to follow a recurring pattern. They often screen long lists of companies, and gather published data on possible acquisition candidates. Then, something else happens: (a) through failure on some aspect of the acquisition, management may find itself unable to complete anything of substance; (b) faulty analysis may lead management to acquire a poor company; or, (c) they may pay so much for a good company that it cannot earn an acceptable return on the investment.

The best approach to acquisition negotiations is a basic plan which enhances the chances of success in merger negotiations. According to MacDougal and Malek, such a plan shows negotiating executives, often perplexed by the maze of details surrounding a prospective acquisition, how to handle three critical aspects of the negotiating process: (1) Laying the groundwork by understanding the financial relationships for any agreement between the two companies. (2) Resolving the financial facts so that an early and realistic agreement on terms is possible. (3) Developing the seller's interest in the nonfinancial aspects of the proposition.

The Groundwork

MacDougal and Malek (1970) found, as the first step toward a successful acquisition, the acquiring company should lay the groundwork in two areas: (a) it needs to think out carefully and express in specific terms its own corporate objectives and acquisition criteria; and (b) it needs to study and understand the basic financial relationships between itself and its potential acquisition.

Objectives.—At a minimum, corporate objectives should include such financial standards as earnings per share, growth rate, and annual return on investment. Without these yardsticks, they report, there is no way of knowing whether a given acquisition would act as a spur or a brake on overall corporate performance. The earnings growth rate (EPS) and return on investment (ROI) that an acquirer should set as its goals depend on many factors that will emerge only from a thoughtful evaluation of internal corporate potential combined with a sound assessment of the business environment. Some companies, MacDougal and Malek found, regularly measure themselves on two counts: first, against their own industry; then, against a cross section of all manufacturing companies, such as that provided by Fortune's "500."

For example, they suggest, an earnings per share (EPS) growth rate of 12% a year compounded would equal the performance of companies at the bottom of the top quartile of Fortune's "500" during the past decade. A 15% return on shareholders' equity would also place a company at the bottom of the top quartile during this period. In any event, the end result should be a goal that is challenging but attainable.

Prenegotiation.—Once the buying company has spotted a promising acquisition candidate, has become acquainted with its principal characteristics, and is preparing for the first approach, the time has come to examine certain key relationships in order to determine what price can be paid—and in what form—without undermining corporate objectives. To determine these relationships, analysts report an analytical approach with graphs that has proven effective in practice. Graphs make it simpler to grasp the essentials before talks begin, as well as provide a sound understanding of the trade-offs involved and the limitations on purchase price and terms.

Finally, the graphs themselves become persuasive tools for the acquiring (A) company's negotiating team. When shown to the seller (S), they help in moving talks on the crucial financial details from an emotional or intuitive level, according to MacDougal and Malek, to a more objective plane where the facts convincingly speak for themselves.

Therefore, in contemplating the purchase of company S, the first task for company A is to project a range of possible compound after-tax EPS growth rates for company S, and then to ask: What is the maximum price we can pay at each of the various company S growth levels if we are going to meet our ROI target?

Figure 14.8 shows graphically this relationship between purchase price, projected company S growth rate, and company A's return on investment. (Consult MacDougal and Malek (1970) which describes in detail the assumptions underlying this exhibit. A financial analyst familiar with basic present-value techniques can generally construct a similar chart.)

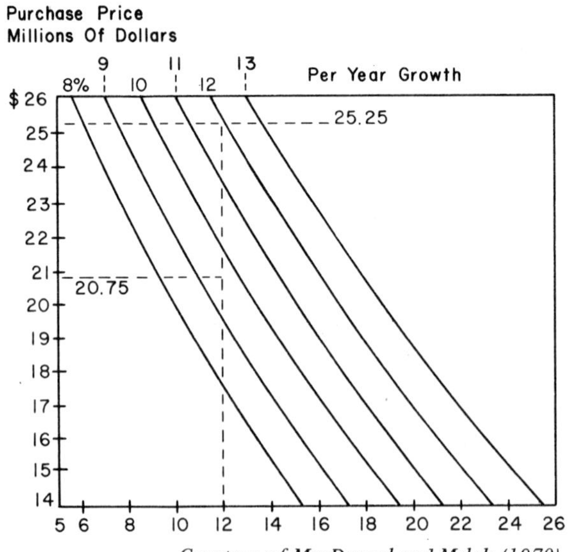

Courtesy of MacDougal and Malek (1970)

FIG. 14.8. COMPANY A'S ROI AT VARIOUS GROWTH RATES AND PURCHASE PRICES

In this example, if an annual company S growth rate of 10% seems probable, then company A can pay $20.75 million for the company and get a 12% return on investment. However, if company S is expected to grow at a rate of 12% annually, a price as high as $25.25 million can be paid without falling below the ROI criterion set by company A's board of directors.

Company A's management, therefore, knows that negotiations on the purchase price should stay within a range of 20-25 million dollars and that the critical factor is the company S projected EPS growth rate.

Seller's Interests

Before negotiations, the experts suggest that the management of company A should develop a preliminary growth plan for the new, combined enterprise (AS) based on careful research into company S's business and industry. Nothing will cause company S to lose interest faster than a naive, superficial growth plan which overlooks important industry factors.

What is suggested to be presented is a thoughtful, fact-founded plan, in writing, which is considered a preliminary suggestion. Executives of company S should be asked to contribute their ideas. This will strengthen their interest in seeing that the negotiations succeed. According to MacDougal and Malek, a combined growth plan might include: (1) A statement of corporate goals for the new enterprise. The combined company is going to achieve something that would not be possible for either alone. (2) A joint marketing program to capitalize on existing economic trends. (3) A tentative program for further acquisitions in the same or allied fields. (4) A thoughtful organization philosophy. This should include the reporting relationships, compensation arrangements, employment contracts, and director representation.

Organizational questions are matters of great concern to the selling company's owners and managers. Often, some are awkward to bring up without appearing more interested in the acquisition than they would like to admit.

Persuading the selling company of the nonfinancial advantages of the acquisition cannot be accomplished through good analysis or charts. The experts suggest that there are at least three steps to take to set company A apart from the crowd and give its offer a genuine competitive advantage: (1) Learn more about the seller's business than other prospective acquirers. (2) Develop a combined growth plan that will capture the imagination of the seller. (3) Develop the sound personal rapport that is prerequisite to a successful working relationship.

WHY DIVERSIFY?

Diversification into new business activity has changed the character of many companies. But, can a company diversify? When should it? Even with sales of a billion dollars a year in a basic business? Some identifiable reasons include the following:

(1) If a controversy has slowed the consumption of products, resulting in a static market condition, and if there is no indication that these pressures will vanish, prudent businessmen will take steps to broaden their base of operations.

(2) If there is a need for greater utilization of financial resources available, diversification into other industries will offer a more productive use of dollars.

(3) If, when dealing with other people, companies, or products, growth is needed, diversification can provide a change that can stimulate a corporation's growth (MacDougal and Malek 1970).

With a decision to do something comes the requirement to choose what to do. This can come from the answer to the question: Can the desired goal of a broadened business base be achieved by developing new technology internally, by acquisition of business knowledge from external sources, or both? Both acquired businesses and internal development are fundamental in a plan for diversification.

ORGANIZATION FOR GROWTH

As a case history, Bright reported, in 1968, about a company that had used over $200 million recently for acquiring companies; it had spent additional money to expand and modernize facilities obtained with these acquisitions to improve their vigor; it also expanded its R&D capability internally to provide support for these acquisitions as well as the existing business. He indicated this was done by building an $8 million product and process development complex which, combined with existing research laboratories, allowed a doubling of R&D effort. With these facts in mind, let us look at the acquisitions and technical activities which resulted in that company.

Growth objectives were set to meet the specific needs of the company and involved internal development plus acquisitions.

In recognition of this, Bright (1968) reported that science and technology were organized for internal growth and closely linked with business analysis, and planning and negotiations for external growth through acquisitions. Guidelines were established for growth areas:

(1) Maintain the existing business by continuing to provide improvements in profits and sales. This can be done by: (A) Lowering costs by finding ways to manufacture present products more effectively without sacrifices in quality. (B) Improving the quality of products offered. (C) Providing new varieties for existing products to reach new groups of consumers in order to hold the present profit base.

(2) Make dramatic new products for the company's existing facilities to reach existing consumers. While this is more speculative and costly than maintenance of an existing product line, it offers great rewards if successful.

(3) Diversify into wholly new areas with a new group of products to enter new markets. This can be done by: (A) The internal development of totally new products. This demands management competence in choosing which profit

center to build—not an easy task. More often than not, internal development to yield a highly desirable profit center is not practical. (B) The acquisition of other companies or products to provide a more expeditious route to a market technology. This would offer a new base on which to build alternative new products or processes coming from a company's own R&D efforts.

While these guidelines were generalities, the company recognized that they had specific strengths that should be mobilized: management had understanding and experience in businesses producing and selling consumer goods; it had manufacturing, distribution, and engineering skills that could aid certain acquired businesses; and, it was well acquainted with the vagaries and uncertainties involved in dealing with agricultural products. All of these then led, according to Bright, to diversify into the food business.

With such resources available, why did they choose foods, rather than electronics, home furnishings, or educational materials, all of which have high rates of predicted growth?

First, Bright reported, corporate skills and strengths were applicable. For example, the marketing people and corporate management had an understanding of the advertising, selling, and distributing know-how necessary for success in the consumer nondurable goods field. Management experience and knowledge were available and useful to the food consumption industry.

Second, investigations revealed that convenience food products were expected to grow at a rate of about 5% per yr for the next 10 yr, not too different from the "glamour" industries. Also, there is a revolution going on in the American kitchen made necessary by the housewife's need to save time and effort. Science and technology are at work meeting this need with new products. These were being built on changes in food preservation, processing, packaging, and distribution.

These factors were recognized by those responsible for corporate growth as areas falling within the guidelines established earlier.

Market exploration techniques applied to this situation, and as discussed earlier, are shown in Fig. 14.9 and 14.10. As can be seen, other alternatives would have been to enter into other agriculturally-based products, including the forest products business, etc.

Active Diversification

Ready to make diversification efforts toward an acquisition in the food industry, they established a food product and process development group in their R&D organization and in marketing and other fields. Two new food processing techniques were chosen for study. A consultant with extensive experience in food marketing, particularly in the planning and test marketing of new products, was hired who acted as a "shadow marketing department" to establish necessary product physical, packaging, and pricing specifications. This provided an instant base in the desired discipline of the marketplace for the

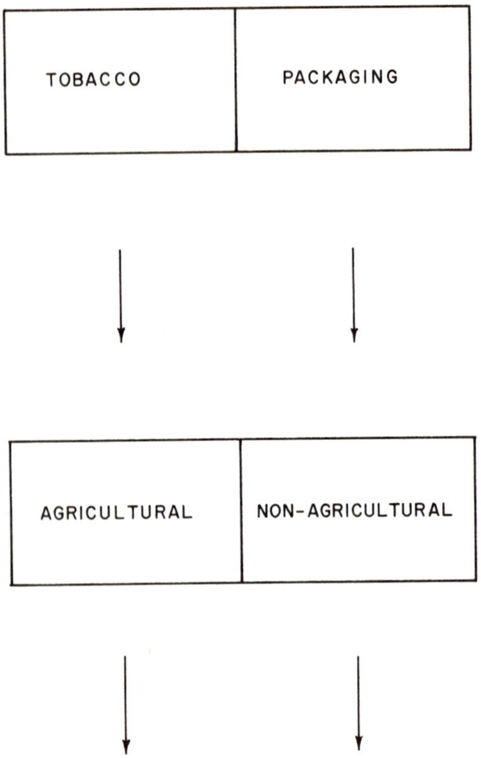

FIG. 14.9. AREAS OF COMPANY BUSINESS

technical group, even before they had an operating company with which to work. It also made a smoother transition possible once the acquisitions were made.

The first acquisition, according to Bright (1968), was made to provide a manufacturing vehicle for results of early laboratory work. Another was acquired which had products in a growing segment of the foods business, but had had few new products introduced in recent years to exploit a group of excellent, existing brand names. In this case, new and different products are being developed at the request of the parent company's management to expand these franchises more effectively.

The science and technology for all new businesses were centralized in their R&D center. This had been done because they believed that the total effectiveness of a combined group would be greater than that of several small units. Also, more scientific disciplines could be applied to a given problem and a wider range of skills and equipment would be available for use.

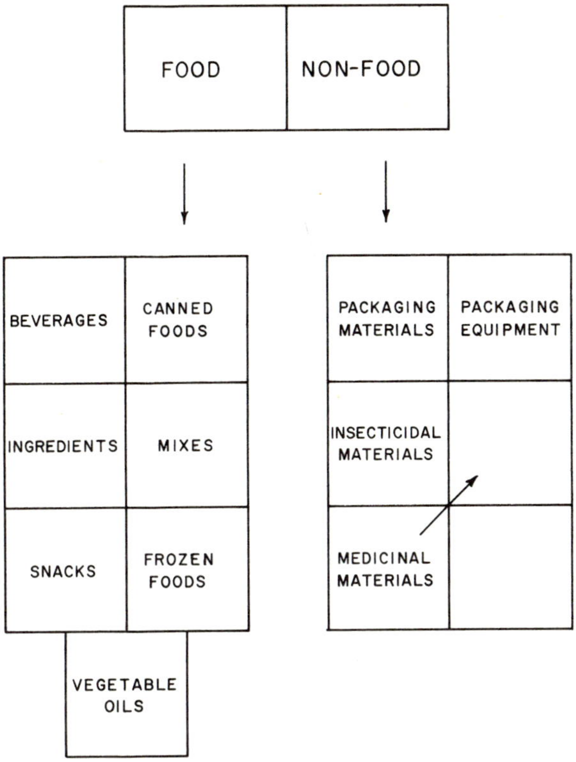

FIG. 14.10. SYSTEMIZATION OF USES

Furthermore, communications in the technology areas would be easier to accomplish. They, as have others, found that it is not too difficult to communicate with subsidiary managements even though their headquarters are not in the same area. Had they chosen to maintain and expand one of the laboratories in an acquired company for their product development work, they would not have maintained the needed close relationship with their packaging development people or basic research groups.

Upon reflection, Bright believed a different path could have been followed had they made a larger acquisition in the food field at the start. As it is—and as others have found—he believes that centralization is the right way to operate science and technological support activities in relation to acquisitions.

In this company's growth area organization, scientific study, business analysis, and economic evaluation of projects were closely joined against the kind of "Merger Lemons" that resulted in failure for 1 out of every 6 mergers involving manufacturing firms in the United States in the 10 yr ending 1965. Too often there has been insufficient appraisal of the scientific and technological trends at work that could and would affect the success of acquisitions.

As a practical experience to be shared, this company avoided some possible mergers by having careful scientific and engineering studies completed before final steps toward an acquisition were taken. In one instance, a patented process was offered as the main asset of a small and apparently growing business. Examination revealed some changes in the process were needed to resolve quality control problems and this would make the product uneconomical. The growth was in a nonrepeat business. They dropped the idea. Other similar experiences in many companies have revealed the value of applying scientific evaluation to acquisition proposals.

Through acquisitions, this company added a wide range of foods, snacks products, and fruit juice beverages to their lineup of leading tobacco brands. During 1967, several of their acquired organizations were consolidated into a subsidiary responsible for manufacturing and marketing all of these products. Bright reported this subsidiary (with sales large enough to rank in Fortune's top 500) had been constructed in 5 yr. Through other subsidiaries acquired, they became suppliers to the food processing and handling industries, as well as being direct participants. They now supply starches and corn sweeteners, meat and produce wraps for supermarket use, packaging for hundreds of millions of pounds of margarine, as well as for a great many other food products.

Both internal growth and acquisitions have played a part in this diversification. A packaging company grew within the parent company as an integrated supplier of cartons and other packaging materials.

This case history provided by Bright reveals that, in order to diversify a business effectively, a program must be constructed on the rationale of corporate strengths to meet the needs of the times and availability of resources. Management's professional judgments play an important role—both in internal development and the acquisition of other companies.

An increasing number of companies are growing more conscious of the opportunities of growth through diversification and acquisitions. The systems and procedures outlined above are useful introductions to the overall subjects.

In the next chapter, we will explore building a business from a product, and, as a corollary, examining the options available in selling a product which does not fulfill long-range plans.

BIBLIOGRAPHY

Adler, L. 1967. Systems approach to marketing. Harvard Business Rev. *45*, No. 3, 105-118.

Anon. 1970. Ethics for Executives. Marketing, public relations and business policy. Harvard Business Rev. Reprint Serv., Soldier's Field, Boston.

Anon. 1970. Finance. Analysis of corporate earnings, stockholder relations, and new techniques. Harvard Business Rev. Reprint Serv., Soldier's Field, Boston.

Anon. 1970. Mergers and acquisitions. Harvard Business Rev. Reprint Serv., Soldier's Field, Boston.

Anon. 1970. Marketing planning and strategy. Harvard Business Rev. Reprint Serv., Soldier's Field, Boston.

Bright, W. M. 1970. Science and mergers and acquisitions in corporate growth. Food Prod. Develop. *2,* No. 4, 43-50.

Cooper, A. C. 1966. Small companies can pioneer new products. Harvard Business Rev. *44,* No. 5, 162-179.

Diehl, R. K. 1969. Market segmentation vs. positioning in product analysis. Food Prod. Develop. *3,* No. 5, 46-64.

Hilton, P. 1967. New Product Development. Prentice-Hall, Englewood Cliffs, N.J.

Jenkins, F. A. 1970. The culmination of product development and market research in consumer communications. Food Technol. *24,* No. 4, 60-67.

Levitt, T. 1965. Industrial purchasing behavior. Harvard Business School, Div. Res., Boston.

Levitt, T. 1966. Innovative imitation. Harvard Business Rev. *44,* No. 5, 63-70.

MacDougal, G. E., and Malek, F. V. 1970. Master plan for merger negotiations. Harvard Business Rev. *48,* No. 2, 71-82.

McDaniel, G. 1964. The meaning of the systems movement to the acceleration and direction of the American economy. Proc. 1964 Systems Engr. Conf., Clapp and Poliak, Inc., New York.

O'Meara, W. H. 1961. Selecting profitable products. Harvard Business Rev. *39,* No. 1, 95-102.

CHAPTER 15

Build a Business or Sell the Product

A new venture group is an entrepreneurial team assigned direct responsibilities for achieving a profitable return from the investment assigned to it. Usually, it is an internal organization to a parent company. The venture technique is being used more and more to help companies grow internally. Experience reveals that, for success, the mission of the venture group should be limited: develop a new product area and turn out profitable products. Being a high risk group, the venture team needs reward opportunities equal to their personal risks. The attractiveness of the proposition to entrepreneurial types of people, and the successes available for review, suggest that it is highly functional. In addition, it is worthy of careful review by companies since it also provides a flow of personnel into middle management levels, important in developing and keeping such people and maintaining corporate morale.

In this chapter, we will explore the nature of venture groups, then expand this to an example of the practical application of the principles involved. Then we will explore the idea of selling unused venture possibilities, once the parent company decides for one reason or another that it will not pursue the area it has developed.

BUILDING WITH VENTURE GROUPS

Bates (1967) stated that the venture approach was better, faster, cheaper. What did he base this on? The waste which occurs in new product development today.

For example, many people think of the process of creating new products as being something like a funnel, where 80-100 product ideas are screened down to 40-50 concepts on which testing will be conducted to get 5 new products into development. Out of the 5, 2 might reach test marketing, and most people would settle for 1 successful new product. Yet as we have seen, the funnel is not an isolated activity but a continuing process, and the quality of new product ideas that go into the top of the funnel determines what comes out at the bottom. There is necessarily waste.

Causes Behind Waste Effort

What causes this high percentage of waste in new product expenditures? According to Bates and others, the problems which cause waste are lodged in every function dealing with new products, with two basic causes: (1) A lack of feasible new product ideas. (2) Inability to move new product projects decisively through organization.

In order to examine what causes new product expenditures to go up, we can begin by looking at the obstacle course of problems to be surmounted by the individuals who influence new product development.

Experience suggests that the new product director often starts with top management's pet projects; develops long lists of product ideas for screening purposes, struggles with research to get "here and now" prototypes; fights for straight answers on product development from technical people; struggles to get sufficient, high quality production in spite of manufacturing's primary involvement with existing products in terms of capacity and efficiency; comes up with exciting new market analyses which identify significant profit opportunities; copes with day-to-day problems that consume a great deal of his time, but do not necessarily contribute to his productivity; finds himself constantly beset with requests for superfluous research from committee management that frequently cannot decide what to do next.

The technical research director often feels he is being given insufficient guidance as to what is wanted; is forced to come up with new product ideas to justify his existence; and tends to spend money on many product ideas that never get off the ground.

The financial director rarely (if ever) is allowed to join the action until it's too late.

Many causes of wasted effort are traceable to both undercontrol and overcontrol. Undercontrol due to a lack of central responsibility for a return on new product investment is often combined with a lack of a common objective for the various functions. As a result, one department tends to work at cross purposes with another. Overcontrol occurs when authority and responsibility are not vested in one individual who can move decisively to get the job done with dispatch.

The Venture Approach

Is there a better way? There must be, and there is. To those using the Venture Approach to new products, it is the only way.

How does a venture team actually work? What does each team member do? Bates (1967) summarized it as follows: If it is a one man team, the man would be the venture manager. Usually it is a group. The functions of the venture team manager include: (1) Top management contact. (2) Concept development. (3) Marketing strategy decisions. (4) Client solicitations (inside the firm to divisions which may be considered as clients, or outside the firm to marketers of a new product). (5) Team administration.

If it's a three-man team, the second man must serve as the marketing/economic/financial representative with responsibilities for: (1) Concept development. (2) Product profitability statements. (3) Market research programs. (4) Marketing planning. (5) Sales testing. (6) Product administration. (7) Venture cost control.

The third man on the team must be a technical/research production representative, responsible for: (1) Concept development. (2) Technical feasibility. (3) Supervision of technical research programs. (4) Product and package development. (5) Co-packer arrangements if needed. (6) Production through test marketing.

The first order of activity for this team is to establish new product objectives. What distribution channels are open to the team? How much gross profit must be developed on a specific product (in both dollars and percentage)? What profit before tax criteria is considered adequate for a new product? What innovation will be tackled: a product the company wants; creation of a new industry; or innovations of products within a specific category?

Having established basic criteria to measure ideas, then the concept phase starts by identifying products to be developed. Generating new product ideas requires disciplined creativity, as we have discussed earlier. Each team member must participate both in a creative sense and a critical sense to assure that concepts will have technical, financial, and consumer benefit targets.

To return to the central entrepreneurial theme of the venture approach, what drives it is the conviction and determination of its team members. One must be able to visualize a new product concept before it becomes an entity. The venture group must focus this kind of vision to project what is to what can be. And, as all experienced with new venture teams know, behind every new product there is a believer.

BUILDING A BUSINESS VENTURE

Recently published data on a new development of a candy type product by a large flour company offers an opportunity to illustrate a venture team at work. Hase and Riter (1970) reported that Space Food Stick, as it was called, actually originated when the National Aeronautics and Space Administration (NASA) came to the company with a request for a complete meal in stick form, to be eaten under the strenuous conditions of manned space flight. Requirements for the product were totally outside existing bounds of company technology. The product was required to have the following characteristics or "must-haves": (A) Provide instant energy. (B) Provide lasting energy. (C) Be nutritionally balanced. (D) Be good tasting. (E) Be portable so that it can be consumed under the exacting conditions of space flight confinement. (F) Be fabricated so as to not produce crumbs, which under weightless conditions would cause problems.

The company subsequently reported that its laboratory had successfully developed a product meeting these criteria. Food scientists were able to combine in a good tasting, convenient form the complete proteins, carbohydrates, and fats—in the proportion necessary and recommended—to sustain life for an indefinite period. This product was produced in such a way as to add nutritional content—that is, more protein—while keeping caloric and bulk content very low. NASA reported it a breakthrough. Hase and Riter indicated that, having developed a technology which could produce a product with all these characteristics, it was natural that they should consider its consumer possibilities. But it was not only the availability of the technology which led them to make this move. The technical ability certainly was an important factor; but alone it was not enough. What led them to consider consumer possibilities were market research findings, not product-oriented at all, but studies undertaken to evaluate changes in eating patterns of American families and the roles various family members played in those patterns. These were also discussed in Chap. 2.

Some of the basic findings of these projects bear reporting. The traditional family three-meal-a-day pattern in American eating is fading. More and more people are getting a large part of their daily food intake via between-meal snacking. These nonmain meal foods are generally high in fat and carbohydrates, and low in protein. The primary preparation of meals is shifting from the mother/wife to the individual family member. Time spent consuming food at the table is decreasing; and item/portion control is being assumed by the individual family member.

Meanwhile, research by government agencies showed that many Americans, despite their affluence, are undernourished. Given both the government studies and their own, they felt that the product developed for NASA could fill consumer needs identified in their basic research on dietary patterns. This became the real reason why the company decided to commercialize the product.

VENTURE TEAM EFFORT

Viewed from the vantage point of the venture team manager, we can identify the events that unfolded from published data.

With appropriate personnel selected and assigned from marketing, production, and R&D, they viewed the scene. As noted earlier, the development of Space Food Sticks was different from other company projects. However, they did have a product for screening. Preliminary product testing revealed product work to be done, and major R&D effort in support of the consumer product was available on call.

The development decision was made following group interview research. In this case because of capital requirements, they waited until they had enough quantitative data to determine whether or not the product idea had strong appeal. From consumer research they concluded that it did.

Next, they had two major efforts to be executed: measure both product development and communication development needed to yield success with the two programs culminating in some measure of the relative compatibility of elements.

The product development process was comprised of five phases. In the first, all product variables that could affect consumer preference ratings were generated. These variables included not only levels of ingredients, but the different types of suppliers of the "same" ingredient (e.g. protein, flavorings), and the types of manufacturing processes that might be used. These variables were then judgmatically screened by the R&D member.

The second phase was to build prototypes for the extremes of many of the variables to be screened. The objective of this second step was to determine which variables had little or no effect on preference. Preliminary screening of these variables was executed by an expert panel.

The third phase was to take the prototype products incorporating the reduced set of variables to the consumer. The nature of these experiments, for example, would test the product with 100% Type 1 protein versus 100% Type 2 protein. If no difference was perceived, the type of protein would be held constant in future product optimization work.

The fourth phase was to determine the best levels for each of these variables in combination such that the preference ratings of consumers were maximized.

All combinations of possible variables were developed into prototypes for consumer testing.

In spite of no clear-cut optimum resulting from these tests, clear direction was found on how to formulate an acceptable consumer product within the constraints of effective manufacturing.

In phase 5, they produced enough of the top 3 prototype products for testing by consumers. The best of these final three was then brought together with the communications development researcher for an in-home positioning and product usage test.

While the product development work had been going on, the four winning positionings were being modified, deleting some of the undesirable factors identified, adding positive elements from others, but mostly enhancing the basic selling theme.

Their next step was to identify as many consumer benefits the product possessed that might be important, and to develop possible ways of presenting these benefits to consumers. Group interviewing continued following this process until they had reduced initial positioning to 12. Each positioning was singular in stress; that is, the position stressed one consumer benefit or attribute. The visuals were used to enhance the theme desired. Research was conducted among a representative sample of men, women, and teenagers. Each respondent evaluated 2 of the 12 positions using a balanced, incomplete block design. Results of the test are shown in Table 15.1.

The 4 positionings, again executed as 4-color full page print ads, were taken door-to-door to a representative sample of housewives, with the product left with her for her family to try. The change in response after trying the product was the measure of how well the product satisfied the promise. The results of this test appear in Table 15.2.

As before, results indicated that position D had the highest initial appeal and was also seen to fulfill the promised benefits.

Upon completion of this test, they felt quite confident of the potential for the

TABLE 15.1

RESPONSE TO INITIAL MARKET POSITIONINGS

Position	Probability of Purchase (%)	Rank
A	12.1	9
B	16.3	7
C	18.3	5
D	36.8	1
E	7.1	10
F	5.6	11
G	14.2	8
H	16.5	6
I	26.5	3
J	22.2	4
K	3.2	12
L	29.6	2

Source: Hase and Riter (1970).

TABLE 15.2

RESPONSE TO VARIOUS SPACE FOOD STICKS WITH POSITIONING BY DEGREE OF EXPOSURE TO PRODUCT

	Probability of Purchase by Positioning			
Stage of Test	D (%)	L (%)	I (%)	J (%)
Before taste	32	26	30	20
After taste	28	22	20	19
After extensive usage	30	20	18	20

Source: Hase and Riter (1970).

product, since a decision to enter the market required major capital commitments. Therefore, they decided to undertake a Buyer Reaction Test. Product is put into final packages with labels (after testing several alternative labels) and demonstrated in grocery stores, encouraging interested consumers to buy. If they do buy, they ask if they might call on them in a week or so to get their response to the product and their intentions to buy again. They needed a valid indication of future repurchase intentions. Buying intention scores from other studies of this type correlate highly with actual market repeat rates.

Hase and Riter (1970) reported that the product met their criteria on the basis of the buyer reaction test; therefore the venture group recommended they enter test market. The first market was opened in January 1968. In the test market both controlled and uncontrolled stores were used. The objective of two independent audits was to provide an estimate of sales in an environment representing something close to reality (uncontrolled stores), and an experiment where one could control everything, such as facings, displays, etc., and measure the effect of different marketing variables.

In test market they marshalled a heavy TV campaign, using the commercial that had been chosen from the earlier in-home test, based on the D positioning of Table 15.2. These commercials had been tested via coupon redemption methods prior to the formal decision to go into that market. As to be expected, they faced many problems in test market, including packaging problems, inability to control prices in all test stores, underestimating consumer demand, production problems creating product shortages, and consumer protests directed at the prevailing grocery product prices in general (Table 15.3).

TABLE 15.3

CONTROLLED TEST DESIGN WITH PRICE PER BOX CONSTANT

	Store Section 1	Store Section 2	Total
Yield 1	8	8	16
Yield 2	8	8	16
Total	16	16	32

Source: Hase and Riter (1970).

However, they were able to determine three things: (1) The product moved. (2) Locating the product in Store Section 2 sold better than Store Section 1. (3) Yield 2 was more profitable than Yield 1.

With these results, they decided to open new markets as well as to run a tightly controlled experiment in 4 other cities in which they would test 2 advertising weights, 3 prices, and 2 store sections–a new one, and the other the winner from the previous test. The design of this test is shown in Table 15.4.

TABLE 15.4

TEST DESIGN FOR TWO LEVELS OF ADVERTISING SUPPORT, TWO IN-STORE LOCATIONS, AND THREE PRICES

City	Advertising Weight	Store Section					
		2	3	2	3	2	3
A	High	5	5	5	5	5	5
B	Low	5	5	5	5	5	5
C	High	5	5	5	5	5	5
D	Low	5	5	5	5	5	5

Source: Hase and Riter (1970).

This test was run for six months and yielded extremely valuable results. Not only were they able to measure the main effects of the treatments, but also discovered some interactions between price and advertising weight, and price and time. The two store sections did not seem to differ in terms of sales. The venture team opened other markets during the latter part of 1968.

In each of these markets they also collected awareness and trial data via telephone surveys. To determine if a bias, such as we were accustomed to in surveys of this type, existed (that is, positive awareness for nonexistent products in familiar categories), they also surveyed the markets before product was introduced.

In tracking awareness over time, they found that the early differences between advertising weights equalized, as shown by Fig. 15.1.

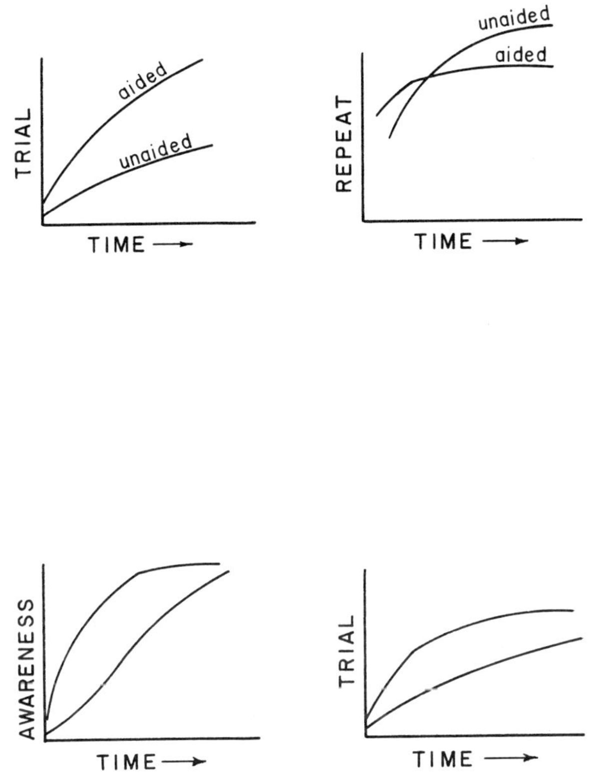

Courtesy of Hase and Riter (1969)

FIG. 15.1. TRACKING AWARENESS

While all this information was being collected, a major capital decision was facing the venture team. They played integral roles in making the financial analyses necessary to evaluate the profitability of the project.

Today they continue to expand distribution and sales are meeting expectations. They are also learning how to appeal to even more customers, and perhaps with other products for a line.

These kinds of results were obtained only because of effective and rigorous execution of the venture team spirit which achieved desired results. The system seems destined for widespread application in new product development.

MARKET UNUSED PRODUCTS AND TECHNOLOGY

What would have happened had the company decided, after all that effort and involvement, that the product did not fit the corporate image or long-term plans?

Hilton (1970) has eloquently stated that the creation of new products is a firmament made up of a host of galaxies in which technology is but one. Members of each galaxy tend to observe the others and interact to them. The degree to which they perform effectively within their own galaxy can, to a large degree, be a reflection of the extent to which they observe what is happening elsewhere in the firmament.

Based on the experiences of many industries, Hilton (1970) proposed several alternatives to the "ordinary outlook" in new food product development:

(A) Upgrade the end product resulting from new product efforts. Ideas and semideveloped products can figure into profit planning. This can be accomplished by employing external outlets for ideas and semideveloped products not pursued internally.

(B) Means should be found, or techniques established, to increase the cross licensing that takes place within the industry. This will tend to spread the benefits of improved technology and, by reason of the revenue it generates, help fund previous and future R&D.

Hilton has found that top management itself, in all industries, has been somewhat remiss in recognizing that technology can and should be licensed—both unused technology that reposes in the corporate files and proven technology which has applications beyond those with which the corporation itself is involved. Every year industry as a whole is wasting countless dollars in potential profits by ignoring the revenue-producing possibilities inherent in exploiting residual technology and patents. This state of affairs is due to the fact that patented technology reposes in the care of the corporate legal department, which, by nature, he claims, regards itself as custodian, rather than exploiter, of such technology.

Hilton suggests the food industry set up Technology Marketing Departments, as other industries have done. These departments should be headed by market-oriented men, he feels.

(C) During the 1970's both R&D and management will, of necessity, be more selective in determining which new products to market. The industry should spin-off to others the ones with marginal interest, not because they have no merit, but because someone else can exploit them better. This, obviously, would call for a cold, clinical evaluation and probably should be made by men who are not emotionally involved with the creation of such products.

(D) The food industry should, as other industries have done, create a new position—a corporate product searcher—to locate spin-offs of fully developed products and technology from other companies in the industry, from other industries, from the government, and from private research sources, according to Hilton (1970). Professional product searchers are being retained by companies in many other fields to augment internal R&D activities with an input of fully-developed products and technology from external sources.

(E) A patent audit should be made of existing technology that resides in patent form in the legal department with a view toward marketing it externally if, upon appraisal, it is decided not to do something with it internally. An estimated 95% of the food processors in the United States have never had a patent audit made that was accompanied by a marketing evaluation of such patents. And, the chances are that 99% have not had such an audit within the past 5 yr (Hilton, 1970).

Some corporations have embarked on an aggressive, well-organized program to market their unused new products and technology. Developments, fully patented, are available for use and exploitation on a licensing basis by other companies.

In addition to detailing the patentable features, the nature of the new product or technology, and the commercial advantages, some companies are now offering reports, accompanied by market evaluations that reflect the potential of the product or technology, for the firm that acquires it.

A program for marketing technology bears careful attention. Firms in widely diverse fields are doing it, each in their own way. There may well be a proliferation of such activities in all industries, including the food industry, in the years ahead. It could maximize returns on investments in new product exploration and result in material advantages to industry and to the economy as a whole.

BIBLIOGRAPHY

Angelus, T. L. 1970. Improving the success ratio in new products. Food Technol. *24,* No. 4, 29-36.

Anon. 1970. Finance. Analysis of corporate earnings, stockholder relations, small business financing and new techniques. Harvard Business Rev. Reprint Serv., Soldier's Field, Boston.

Bates, B. 1967. Venture group approach to new products. Food Prod. Develop. *1,* No. 5, 34-42.

Carson, G. D. 1969. Product development decisions using Bayesian decision theory. Food Prod. Develop. *2,* No. 6, 46-50.

Cavalier, P. A. 1968. The financial function in product development. Food Prod. Develop. *1,* No. 6, 28-30.

Cavalier, P. A. 1968. The Balance Sheet. Food Prod. Develop. *2,* No. 1, 32-34.

Goodman, S. R. 1970. Using decision guides for research and development. Food Technol. *24,* No. 4, 42-47.

Greyser, S. A. 1964. The case of unproductive products. Harvard Business

Rev. *42,* No. 4, 38-47.
Hase, P. F., and Riter C. B. 1970. Building a technology into a business. Food Prod. Develop. *4,* No. 2, 16-20, 64.
Hilton, P. 1970. Unexploited sources of new product opportunities. Meeting in Miniature. N.Y. Inst. of Food Technol., Sheraton Hotel, New York. Jan. 20, 1970.
Sampson, R. T. 1969. Sense and sensitivity in pricing. Harvard Business Rev. *42,* No. 6, 45-49.
Scheuble, P. A. 1964. ROI for new product planning. Harvard Business Rev. *42,* No. 6, 118-128.
Steiner, G. A. 1970. Rise of the corporate planner. Harvard Business Rev. *48,* No. 5,, 133-138.

CHAPTER 16

The Business of Change

As we stated at the outset, the business of the food industry is a business of changing markets, and a business of changing markets is the arena for new products. The arena is best seen in overview, such as depicted by Adler (1967) and presented in Fig. 5.1, Chap. 5.

It is obvious that organization is required to cope with the situation. But, not just organization . . . we need enlightened organization.

MASS MARKETS-MASS CONSUMPTION

It was not many years ago that the food industry was composed of a vast number of small companies, each producing its own specialty lines for a small geographical market. The revolution in food packaging and consumer preference for ready-to-use, conveniently packaged foods, according to Kable (1970), was slow in becoming apparent to the industry. While there are several examples of manufacturing and marketing food products with techniques borrowed here and there from other industries (one involved the soft drink industry), some products became household names through the application of sophisticated marketing techniques. Kable looked for a common denominator in the achievement of success: one important factor was the emphasis on volume.

Sales Volume

The Haase and Seawell (1970) analysis discussed earlier reveals the trend toward volume as unmistakable. In new food products, volume is a paramount goal; and the only way in which volume can be achieved is through national advertising and selling. This philosophy was the basis used by the early pioneers in the industry, Kable (1970) found. It was used by other manufacturers, and it is here to stay. Success in launching and creating a demand for new food products depends upon generating volume production and sales. The benefits obtained from know-how and volume production techniques appear to him to be the major reasons why the food industry in various countries associates itself with global organizations. In the introduction of new products, he suggests that local manufacturers cannot compete effectively except in rare instances.

The trend toward volume has brought with it numerous management and financial problems. In years gone by, food processors have been involved with the problems of volume. Yet, decades ago, independent retailers banded together in order to obtain the advantages of volume buying and cooperative advertising. Food retailers have set the pattern for the rest of the food industry, and food manufacturers have been somewhat slower to grasp the ramifications of volume production and sales distribution. The business of new products will become more and more a problem of volume production and selling. Kable predicts the industry will become increasingly concentrated in the hands of fewer and larger corporations unless some dramatic change occurs which cannot be foreseen at this time.

Distribution Costs

With the tendency for retailers to become organized in national units, the first problem food manufacturers face in launching a new product is the problem of national distribution. Unlike retailers, most manufacturing companies tend to be less organized nationally. However, with volume capabilities as the goal, it is inevitable that manufacturing facilities will become larger and more organized in more than one geographic location. Products must be distributed throughout the length and breadth of the country in order to take full advantage of the economies of high volume production. Transportation and distribution costs, therefore, become a major expenditure for all manufacturing organizations, and especially so in the business of new products. Thus, the control of transportation and distribution costs is one of the most important factors influencing the future of the food industry.

In addition to the rising costs of distribution, there has been a tendency for labor costs to rise, and this necessitates that production costs fall or the industry will collapse. The introduction of new products must take this into account. The food industry is very competitive and has an enviable record in consumer prices compared to most other industries. The reason for this in the past, Kable suggests, was the factor of volume. Increased production has allowed cost savings to be passed on to the retailer and to the consumer. The problem in this is that reduced prices to the retailer and manufacturer inevitably lead to lower margins which, in turn, lead to lower profits. The profit margin can be increased with an even higher volume of production and sales, but, there are obvious limits here. The limits can often be broken by the successful introduction of new products. This trend is going to continue. Further, the control of costs and overheads is going to become more important in management than it has been in the past, and this includes new product efforts as well. This is not to suggest that the food industry is not cost-conscious. Yet, it is amazing how many more efficiencies can be introduced into a company, or even in government, when increased costs have to be covered with existing funds.

Budget Control

So far we have suggested that the business of new food products involves a problem of producing an ever-increasing volume of goods from fewer and larger organizations. To be effective, companies which are currently in the industry must grow at a rate in excess of the growth of the industry overall. Growth such as this poses serious questions, as revealed by MacDougal and Malek (1970) and others earlier. There are many pitfalls which must be avoided in a fast-growing organization. It is easy to create potential pitfalls and complicated to eliminate them. Such problems can be solved before a new product is launched with adequately designed management and financial controls.

A high level of sales is vital in a new product if the well-being and future growth of the organization is to be maintained. The difficulty in control hinges around realistic budgets. Unfortunately, in the business of new food products it is difficult at best to estimate the impact which a new product will make. The faster a company grows, the more sophisticated must be the methods used, as we saw in Chap. 8.

The setting of a budget for a new product in an industry as competitive as the food industry is not easy. Yet, a budget there must be if any semblance of financial control is to be implemented. Other problems associated with the business of new food products include competitor action. Opportunity may be lost by delay. Similarly, the higher the volume of sales achieved, the heavier the pressure becomes on company resources for the purchase of additional equipment for production.

R&D Efforts

A company must have an adequate R&D organization. It must protect present product profits and secure future profit opportunities through new products. The tasks are great. So are the responsibilities of the production organization to produce and the marketing organization to market profitably. It usually happens that there is a competition for available resources and efforts. Just when profits might be boosted through the introduction of a new product, the R&D organization often discovers some new technique or process which enables an existing product to be produced at a better profit. Both situations are irresistible to a company. An R&D effort is needed for both, in balance with resources. In arriving at this balance, none of us can afford to underestimate the nature of the technological capabilities of high morale research laboratories. As Kable (1970) has suggested, they have a very good track record in this regard.

There is a tendency in the food industry for one company to acquire another company successful with one or more unique product lines. Movements in this direction become more and more difficult as a company reaches a certain size, especially in view of governmental regulations. In many cases in the past, this seemed less expensive than paying directly for the research, development, production, marketing, advertising, and distribution costs associated with the launching of a similar product. Yet, this too must end. Economic life is becoming too sophisticated to have ready solutions for all companies.

THE NOOSPHERE

Chardin (1955), in discussing his views of the phenomenon of man, called attention to the "spheres" of the Earth. There is the physical sphere, the fixed resources of the earth which we see; a biological sphere of living organisms, of which we are a part; and what he called the noosphere, that which results from the application of human intelligence—increasing with increasing human populations and growing knowledge. The noosphere is to be seen and recognized in terms of its components—the computers, communications systems, arts and sciences, cities, and transportation systems on land, sea, and air, etc.

The noosphere is by nature also growing, and this growth is to be shared by our industry as well.

It is no longer the idea of a limited number of jobs for a limited few people. The new age unfolding is that seen in the noosphere; and it is more a matter of recognizing it and contributing to its growth than being controlled by it.

At the same time, we all recognize that there are two sides to this growth: one is to advance the quality of human living; and the other is to have it in a favorable balance with the membership in the human population.

As McLuhan (1964) has noted, our world has changed; we now live in a global village. The noosphere has developed already to the level that what happens in one sector of it is felt in the others. Global television will add the dimension of sight as well to such happenings.

With global television already operating, it will not be long before global advertising will occur. And, we might ask, who can predict what true global marketing will be like? In fact, what could be advertised globally today? Tomorrow? What products are, in fact, available for such an undertaking?

Adler (1967) found a relevant and interesting reference, dating back nearly 800 yr. A monk wrote the following about St. Godric, a merchant later turned hermit:

"He laboured not only as a merchant but also as a shipman . . . to Denmark, Flanders, and Scotland; in which lands he found certain rare, and therefore more precious, wares, which he carried to other parts wherein he knew them to be least familiar, and coveted by the inhabitants beyond the price of gold itself, wherefore he exchanged these wares for others coveted by men of other lands . . ."

How St. Godric "knew" about his markets in the decade of the 1170's we are not told. How some of us today "know" is sometimes, Adler suggests, no less mysterious than it was eight centuries ago. But we are trying to change that; and he hazards the forecast that the era of "by guess and by gosh" is drawing to a close, spurred on by a number of powerful developments. He described them briefly:

(1) The growing complexity of technology and the accelerating pace of technological innovation.

(2) The advent of the computer, making possible analysis of the relationships between systems and their components.

(3) The intensification of competition, lent impetus by the extraordinary ve-

locity of new product development and the tendency of diversification to thrust everybody into everybody else's business.

(4) The preference of buyers for purchasing from as few sources as possible, thereby avoiding the problems of assembling bits and pieces themselves and achieving greater reliability, economy, and convenience.

Not to be overlooked, in this connection, is the tremendous influence of governments in their buying systems for their programs and for their citizens.

At the moment, we are still too close to St. Godric's methods. By pressing the issue, it should help us to confront more intelligently the awesome complexity of new product development, to deal with the opportunities of technological change, and to survive and prosper from the intensification of competition.

Rising Tide of Consumerism

All that has been presented thus far would be lacking in vision if it did not point further to the rise in consumer concern over the safety and nutritional adequacy of foods.

What Are the Consumer Needs in Terms of Corporate Interests?

The consumer demands appeal, convenience, value, nutrition, and safety in all foods, as an everyday occurrence which would allow eating to be more a social event than an exercise in nutrition.

Meeting the challenge of the 1970's then is going to be one of recognizing new scientific discoveries, what is being dictated by consumers in terms of their needs and wants, the matching of these needs to a corporation's interests, and planning to achieve what has to be an overwhelming opportunity for all.

SUMMARY

In the end, it is the corporation involved in mass marketing consumer products which must pay for development from profits earned from sales of products. What is the success path? Let us assume, as Jenkins (1970) suggested, that we have now developed to the point where the product development men, acting in their free creative atmosphere, are producing a veritable flood of products for which the enlightened market researchers are identifying needs, desires, and markets. All have locked arms, marching forward together singing the "Battle Hymn of the Republic." What does the successful corporation do then? Jenkins believes it is the following:

Speak to the Consumer

First, the product speaks. Everything about a product as it lives in the marketplace, he says, speaks to its customer or prospective customer: the product itself—in terms of appearance and demonstrable performance; the package—its design and inherent appeal or lack of it; its price-value relationship to the user;

the reputation of the manufacturer; and the promotion and advertising messages which the prospective customer receives through any and all media to which she may be exposed. The product speaks, willingly or unwillingly, through all of these channels.

Second, it must speak effectively when it does speak, for to do otherwise is a costly waste. This is the point at which product development and market research come together—to speak, to communicate effectively. Jenkins reports that all effective total marketing campaigns in all of their aspects are built around a central idea or theme, such as the "Universal Selling Proposition," also called the "Purchase Proposition." The purchase proposition is that idea which, when transmitted from the maker to the consumer, will produce the maximum sales for the product or service.

His experiences reveal that the great marketing campaigns are usually the result of the same kind of channeled creativity and developed in much the same way as product development people work. Therefore, by supplying the item with a point of difference, a competitive edge, he says, provides the raw material which is then translated into the purchase proposition, which, in turn, yields the truly effective selling campaign.

Jenkins (1970) indicates that if product development is the genesis of the purchase proposition (in a highly oversimplified way) and market research is the genesis of the media plan (which is essentially the selection of the vehicles to carry the message) then media can be selected (within certain broad tolerances) which will deliver that message to the preselected market. Hence, the two functions of product development and marketing come together to join message and media to culminate in the ultimate consumer communication. With this approach, the product can speak and speak effectively.

Listen to the Consumer

Next, Jenkins finds that corporations listen. If they are sensitive and attuned to the marketplace, it speaks back in many and varied ways. Listen to what it says through the most elementary marketing communication feedback—that of raw sales results. Measure share of market. Screen consumer letters of compliment and complaint. Be attentive to the reaction of the trade. Finally, conduct well-conceived market research among both tryers and nontryers of the product. In short, we should listen.

Modify Product, Media, Message

Finally, he cautions, after the corporation has spoken and carefully listened, it modifies . . . not on the basis of what one thinks or what is expedient for the manufacturer at the moment, but on the basis of what the market reveals. Then, modify the product, the media, and the message, as needed.

Ideally, it is the ever-repeating cycle of marketing through consumer communications. Speaking, listening, and modifying. Jenkins (1970) found they embody the contradictory term of never-ending culmination—the culmination of creativity in product development and working in concert with competent market research to produce informative and persuasive communications directed to properly identified consumer groups. He believes these are the keys to the continuing success of all mass marketing consumer-oriented companies. If we can gain the competitive edge through channeled creativity, minimize our risks through extensive market research in competent hands, recognize its limitations, and, finally, speak, listen, and modify, we will be in harmony with all that can be learned from the experience of others who have had outstanding success for decade after decade.

THE FUTURE CHALLENGE

As was suggested at the outset, a business organization has many of the features of all other living systems. Further, the environment within which all living systems exist is itself changing, as it always has, and, to the best of anyone's knowledge, always will.

Change is the way of life for all living enterprises. How the enterprise reacts to the changing stimuli it receives is evidence of its health and well-being.

Neither change nor adjustment take place overnight. They are necessarily slow moving. Organizations that adapt themselves, their products, and their way of doing business to changing times stand the best chance to succeed in the long-run. And, it is no news that a company, as with a person, that does not become a part of the solution becomes a part of the problem.

As was discussed in Chap. 1, communications only uncover at their outset the differences in opinions and ideas held by men. What is needed is dialogue—open-minded dialogue—between the members of management and employees of the company themselves, and with their customers, their suppliers, and their shareholders. If we trust the notion that the common good results from free wills pursuing the truth, we have the bedrock for decisions—decisions commonly reached, which add to the body of common sense held by them all. And, it takes time for seeds, once planted, to grow and develop and produce tenfold or a hundredfold the number planted and nourished.

It is also abundantly clear that the future belongs to those who plan for it, prepare for it, proceed with it, and pursue it.

The future growth of our industry is assured when all recognize that challenge. It reveals more and more opportunities and rewards for all involved. Such is the nature of men and the organizations and activities they sponsor.

BIBLIOGRAPHY

Adler, L. 1967. Systems approach to marketing. Harvard Business Rev. *45*, No. 3, 105-118.

de Chardin, P. T. 1955. Le Phenomenon Human. Editions du Seuil, Paris.

English Edition, The Phenomenon of Man, available from Harper & Row, New York.

Drucker, P. F. 1968. The Age of Discontinuity. Harper & Row, New York.

Haase, A. E., and Seawell, E. R. 1970. Merchandizing. Nabisco's space management program. Snack Foods *59,* No. 2, 48-52.

Jenkins, F. A. 1970. Culmination of product development and market research in consumer communications. Food Technol. *24,* No. 4, 60-67.

Kable, J. C. 1970. The business of new food products. Food Technol. in Australia *22,* No. 3, 110-114.

Lonergan, B. F. 1957. Insight. Understanding what it means to understand. Philosophical Library, New York.

MacDougal, G. E., and Malek, F. V. 1970. Master plan for merger negotiations. Harvard Business Rev. *48,* No. 2, 71-82.

McLuhan, M. 1964. Understanding Media. McGraw-Hill Book Co., New York.

Servan-Schreiber, J. J. 1969. The American Challenge. Atheneum Publishers, New York.

Index